MEI structured mathematics

Pure Mathematics 5

TERRY HEARD
DAVID MARTIN

Series Editor: Roger Porkess

MEI Structured Mathematics is supported by industry:
BNFL, Casio, Esso, GEC, Intercity, JCB, Lucas, The National Grid Company, Sharp, Texas Instruments, Thorn EMI

Acknowledgements

The authors and publishers would like to thank the following companies, institutions and individuals who have given permission to reproduce copyright material. The publishers will be happy to make arrangements with any copyright holders whom it has not been possible to contact.

The illustrations were drawn by Ken Ovington of Precision Art.

Photographs:

Dan Addleman (69); Martin Dohrn/Science Photo Library (90); Claude Nuridsany and Marie Perennou/Science Photo Library (24); Mr K.C.D. Shuttleworth/Bupa Hospitals Ltd. for St Thomas, Lithotriptier Centre (114); Mathematical Snapshots by H. Steinhaus New edition 1960 © 1950 OUP New York (121).

British Library Cataloguing in Publication Data

Heard, Terry
Pure mathematics 5. – (MEI structured mathematics)
1.Mathematics 2.Mathematics – Problems, exercises, etc.
I.Title II.Martin, David III.Mathematics in Education and Industry
510

ISBN 0 340 64771 X

First published 1996

Impression number	10 9 8 7 6 5 4 3 2 1
Year	2000 1999 1998 1997 1996

Typeset by Multiplex Techniques Ltd.
Printed in Great Britain for Hodder & Stoughton Educational, a division of Hodder Headline PLC, 338 Euston Road, London NW1 3BH by Bath Press

MEI Structured Mathematics

Mathematics is not only a beautiful and exciting subject in its own right but also one that underpins many other branches of learning. It is consequently fundamental to the success of a modern economy.

MEI Structured Mathematics is designed to increase substantially the number of people taking the subject post-GCSE, by making it accessible, interesting and relevant to a wide range of students.

It is a credit accumulation scheme based on 45 hour components which may be taken individually or aggregated to give:

3 Components	AS Mathematics
6 Components	A Level Mathematics
9 Components	A Level Mathematics + AS Further Mathematics
12 Components	A Level Mathematics + A Level Further Mathematics

Components may alternatively be combined to give other A or AS certifications (in Statistics, for example) or they may be used to obtain credit towards other types of qualification.

The course is examined by the Oxford and Cambridge Schools Examination Board, with examinations held in January and June each year.

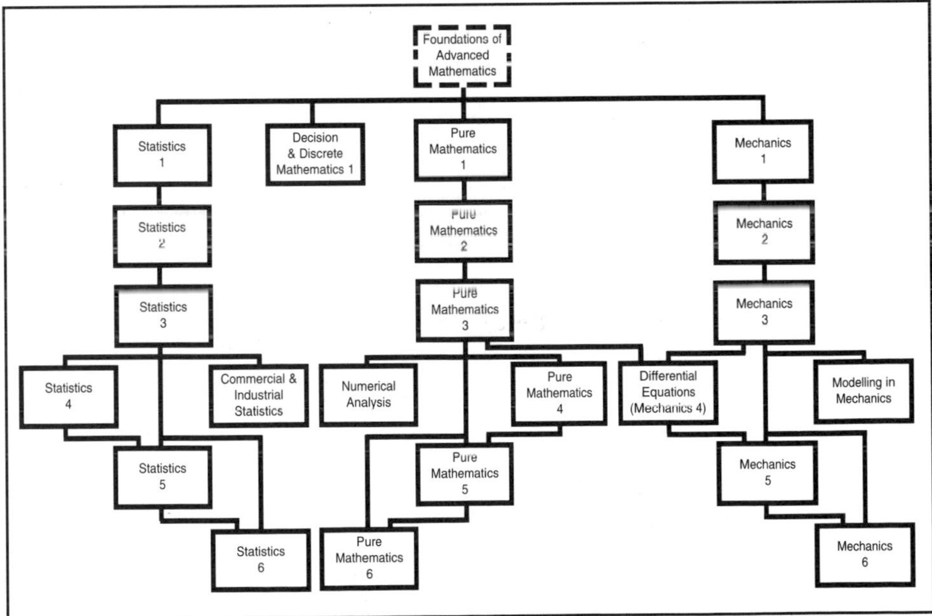

This is one of the series of books written to support the course. Its position within the whole scheme can be seen in the diagram above.

Mathematics in Education and Industry is a curriculum development body which aims to promote the links between Education and Industry in Mathematics, and to produce relevant examination and teaching syllabuses and support material. Since its foundation in the 1960s, MEI has provided syllabuses for GCSE (or O Level), Additional Mathematics and A Level.

For more information about MEI Structured Mathematics or other syllabuses and materials, write to MEI Office, 11 Market Street, Bradford-on-Avon BA15 1LL.

Introduction

This book, the fifth in the series covering the Pure Mathematics Components of the MEI Structured Mathematics course, takes you well into Further Mathematics territory. This does not mean that the new ideas presented here are necessarily difficult, but at this level it is appropriate to work on the basis of reasonably reliable technique (so you should be prepared to complete the details for yourself in some of the worked examples), and to be clear about proving results (or, occasionally, saying that we cannot give a proof yet). For example, in the Algebra chapter care is taken to justify some important results which so far you may have taken as 'obvious'. The Calculus and Complex Numbers chapters build upon the work of earlier components, and links with both these form an important part of the Hyperbolic Functions chapter. There is considerable emphasis on geometry in this book, most obviously in Geometry with Polar Co-ordinates and in Conics, but also in Complex Numbers. The importance of geometry has been somewhat neglected in recent years, and we hope that the full treatment given here will enable you to feel at home in this rich and powerful field.

As always we are indebted to family, friends, colleagues and pupils (not mutually exclusive categories!) particularly Diana Cowey, Ray Dunnett and Mike Jones, for their help in many ways during the writing and preparation of this book. It is a pleasure to record our thanks to them all.

Terry Heard and David Martin

Contents

1 Algebra

In mathematics it is new ways of looking at old things that seem to be the most prolific sources of far-reaching discoveries.

Eric Temple Bell, 1951

For Discussion

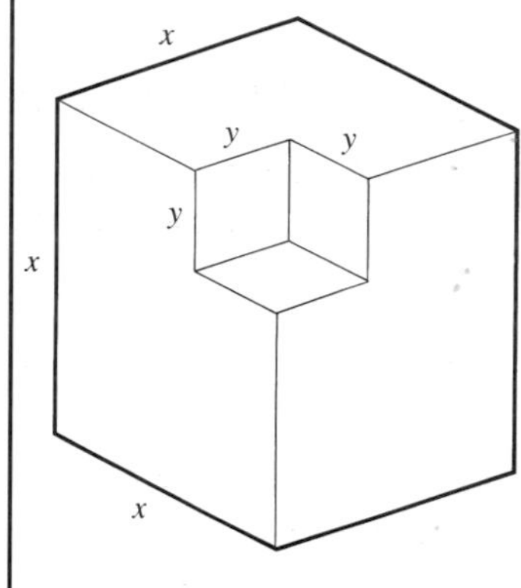

Figure 1.1a

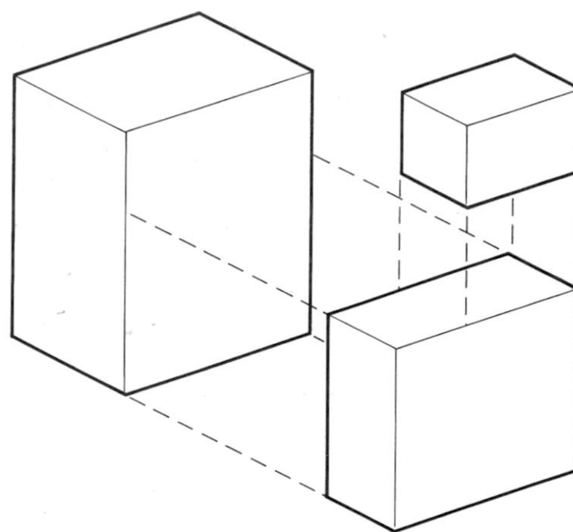

Figure 1.1b

Figure 1.1a shows a cube of side x from which another cube of side y has been removed. Figure 1.1b is an exploded view of the same solid. Explain how it illustrates that $x^3 - y^3 \equiv (x - y)(x^2 + xy + y^2)$.

Identities

During your work in mathematics you will have seen and used statements (known as *identities*) such as the one above as well as the following.

1. $-(1 - x) \equiv x - 1$

2. $(x + y)^2 \equiv x^2 + 2xy + y^2$

3. $\dfrac{a^2 - b^2}{a + b} \equiv a - b$

The symbol ≡ means 'is identically equal to'; it is used to emphasise that these statements are true for all values of the variable or variables for which the functions involved are defined. Statements 1 and 2 are true for all values of x and y. Statement 3 is true for all values of a and b, provided $a + b \neq 0$.

Distinguish carefully between equations and identities.

- All possible values of the variable (or variables) will satisfy an identity, provided only that they are in the domains of all the relevant functions. The identity $(x - 1)^2 \equiv x^2 - 2x + 1$ (for example) is true for all values of x. It does not make sense to try to solve something you recognise as an identity.
- There will be some values of the variable (or variables) which do not satisfy an equation. The equation $x^2 - 7x + 12 = 0$ (for example) is only satisfied by $x = 3$ or 4.

Whenever we use the symbol ≡ you can be sure we are using an identity, but some statements which look like equations (using the symbol =) are in fact identities.

You will also be familiar with trigonometrical identities such as

$$\sin(-\theta) \equiv -\sin\theta \text{ (true for all values of } \theta)$$

and $\sec^2\theta \equiv \tan^2\theta + 1$ (true for all values of θ for which $\tan\theta$ and $\sec\theta$ are defined)

but in this chapter we only look at polynomial identities (those identities in which the only functions used are polynomial functions) and identities that can be derived from them.

Polynomials

Much of the following terminology will probably be familiar. A *polynomial of degree n* is any expression which can be put in the form

$$c_n x^n + c_{n-1}x^{n-1} + c_{n-2}x^{n-2} + \dots + c_1x + c_0 \quad \text{where } c_n \neq 0.$$

Each term is the product of a *coefficient* (c_r) and x^r where r is a positive integer or zero. This may also be described as a polynomial *of order n*; c_nx^n is known as the *leading term*; c_n, the coefficient of the leading term, is known as the *leading coefficient*; c_0, the coefficient of x^0, is known as the *constant term*.

The *zero polynomial* is the polynomial which has all its coefficients equal to 0. Notice the distinction between a polynomial of degree zero and the zero polynomial:

- a polynomial of degree zero has all its coefficients equal to 0 except the constant term: it is also known as a *constant polynomial*;
- the zero polynomial has all its coefficients equal to 0: its degree is undefined.

Adding, subtracting and multiplying polynomials will be familiar. For example, if

$P(x) \equiv 3x^2 + 4x - 5$ and $Q(x) \equiv 2x^2 + 4x - 3$ then

$$P(x) + Q(x) \equiv 5x^2 + 8x - 8;$$

$$P(x) - Q(x) \equiv x^2 - 2;$$

$$P(x)Q(x) \equiv (3x^2 + 4x - 5)(2x^2 + 4x - 3) \equiv 6x^4 + 20x^3 - 3x^2 - 32x + 15.$$

Activity

P(x) and Q(x) are polynomials of degree m and n respectively.

(i) If $m \neq n$ what can you say about the degree of (a) $P(x) + Q(x)$?

(b) $P(x) - Q(x)$?

(c) $P(x)Q(x)$?

(ii) Repeat (i) when $m = n$.

You will have noticed that the product of a polynomial of degree m with a polynomial of degree n is a polynomial of degree $m + n$. This property holds even when one (or both) of the polynomials is of degree 0; this property would not hold if we were to attach any finite degree to the zero polynomial.

Polynomial division

Consider the following division process in which we divide the polynomial $P(x) \equiv x^3 - 4x^2 + 7x - 12$ by $x - 2$.

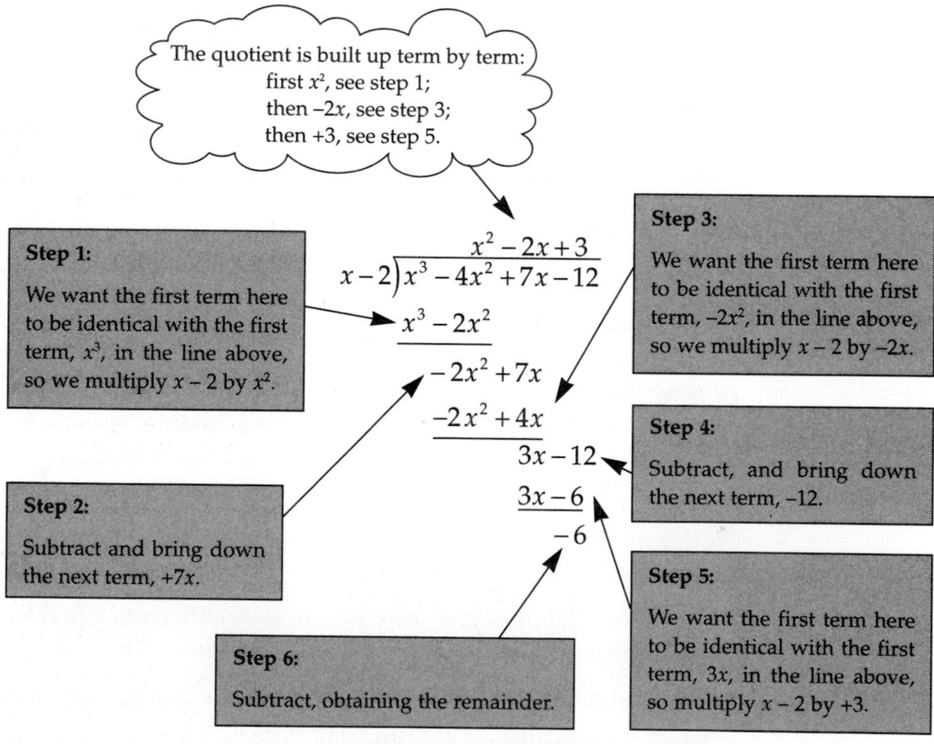

The division tells us that when $P(x) \equiv x^3 - 4x^2 + 7x - 12$ (known as the *dividend*) is divided by $x - 2$ (known as the *divisor*) the quotient is $x^2 - 2x + 3$, and the remainder is -6. We may write this result as the identity

$$\frac{x^3 - 4x^2 + 7x - 12}{x - 2} \equiv x^2 - 2x + 3 + \frac{-6}{x - 2}. \quad ①$$

However it is probably better to write it as

$$P(x) \equiv x^3 - 4x^2 + 7x - 12 \equiv (x - 2)(x^2 - 2x + 3) - 6 \qquad ②$$

since identity ① is valid for all x except $x = 2$ while ② is valid for all values of x. Note the pattern familiar from the division of numbers:

$$\text{dividend} = \text{divisor} \times \text{quotient} + \text{remainder.}$$

Notice that $P(2) = 8 - 16 + 14 - 12 = -6 =$ the remainder. This is no coincidence, as substituting $x = 2$ in ② will show. This property forms the subject of the Remainder Theorem which we state and prove on page 6.

Division by a quadratic or cubic polynomial is performed in much the same way as division by a linear polynomial. We illustrate this in the following example.

EXAMPLE

Find the quotient and remainder when $6x^4 - 32x^2 + 7$ is divided by $3x^2 + 3x - 1$ and write out the corresponding identity.

Solution

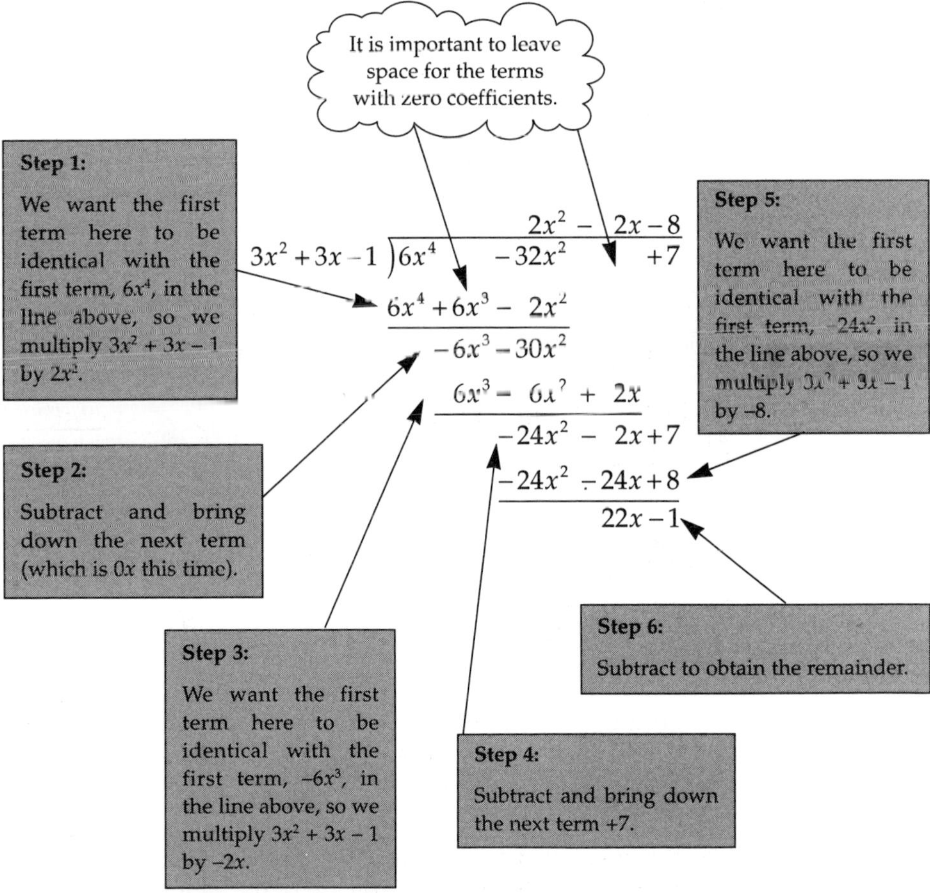

The quotient is $2x^2 - 2x - 8$; the remainder is $22x - 1$.

The identity is $6x^4 - 32x^2 + 7 \equiv (3x^2 + 3x - 1)(2x^2 - 2x - 8) + 22x - 1$.

The Remainder Theorem

The *Factor Theorem* will be familiar (see *Pure Mathematics 1*, page 138). It states that if P(x) is a polynomial then

$$P(a) = 0 \Leftrightarrow (x - a) \text{ is a factor of } P(x).$$

So if we wish to know if $(x - 3)$ is a factor of $P(x) \equiv x^3 - 4x^2 + 7x - 12$ we need only evaluate P(3): $P(3) = 3^3 - 4 \times 3^2 + 7 \times 3 - 21 = 27 - 36 + 21 - 12 = 0$

$$\Rightarrow (x - 3) \text{ is a factor of } P(x).$$

Evaluating P(2) gives

$P(2) = 8 - 16 + 14 - 12 = -6 \neq 0 \Rightarrow (x - 2)$ is not a factor of $P(x)$.

We now extend this result.

The Remainder Theorem

If polynomial $P(x)$ is divided by $x - a$ the remainder is $P(a)$.

Proof

When dividing a polynomial by a linear expression the remainder is clearly a constant (possibly 0). Suppose dividing $P(x)$ by $(x - a)$ gives quotient $Q(x)$ and remainder R: i.e.

$$P(x) \equiv (x - a)Q(x) + R.$$

Putting $x = a$ gives $\quad P(a) = 0 \times Q(a) + R \quad$ so that $R = P(a)$.

EXAMPLE

Find the remainder when $P(x) \equiv 3x^4 - 5x^3 + 2x^2 - 7x + 2$ is divided by

(i) $x + 2$; (ii) $2x - 1$.

Solution

Evaluate $P(x)$ at $x = -2$, the value which would make divisor = 0.

(i) $P(x) \equiv (x + 2)Q_1(x) + R_1$;

putting $x = -2$: $R_1 = P(-2) = 3(-2)^4 - 5(-2)^3 + 2(-2)^2 - 7(-2) + 2 = 112$.

When $P(x)$ is divided by $x + 2$ the remainder is 112.

Evaluate $P(x)$ at $x = \frac{1}{2}$, the value which would make divisor = 0.

(ii) $P(x) \equiv (2x - 1)Q_2(x) + R_2$;

putting $x = \frac{1}{2}$: $R_2 = P(\frac{1}{2}) = 3(\frac{1}{2})^4 - 5(\frac{1}{2})^3 + 2(\frac{1}{2})^2 - 7(\frac{1}{2}) + 2 = -\frac{23}{16}$.

When $P(x)$ is divided by $2x - 1$ the remainder is $-\frac{23}{16}$.

The Factor Theorem is a special case of the Remainder Theorem: if $P(x)$ is a polynomial and $P(a) = 0$, the remainder when $P(x)$ is divided by $x - a$ is 0, and $x - a$ is a factor of polynomial $P(x)$.

We do not have a generalised version of the Remainder Theorem which tells us the remainder when we are dividing by a polynomial of degree 2 or more.

EXAMPLE

When $P(x) \equiv x^3 + bx^2 + cx + 5$ is divided by $x - 2$ the remainder is 3. When $P(x)$ is divided by $x + 3$ the remainder is -67. Find b and c.

Solution

$P(2) = 3 \quad \Rightarrow 8 + 4b + 2c + 5 = 3 \quad \Rightarrow 4b + 2c = -10$ ①

$P(-3) = -67 \quad \Rightarrow -27 + 9b - 3c + 5 = -67 \quad \Rightarrow 9b - 3c = -45$ ②

It is possible to form equations ① and ② by polynomial division, but this takes much longer and there is more room for blunders.

Solving ① and ② simultaneously: ① $\Rightarrow 2b + c = -5$

② $\Rightarrow 3b - c = -15$

Adding gives: $5b = -20$

so that $b = -4$ and $c = 3$.

Nested multiplication

A convenient way of evaluating a polynomial, known as *nested multiplication*, is based on the identity $ax^3 + bx^2 + cx + d \equiv ((ax + b)x + c)x + d$.

For example if $P(x) \equiv 2x^3 + 5x^2 + 7x + 5$ then $P(3) = ((2 \times 3 + 5) \times 3 + 7) \times 3 + 5 = 125$.

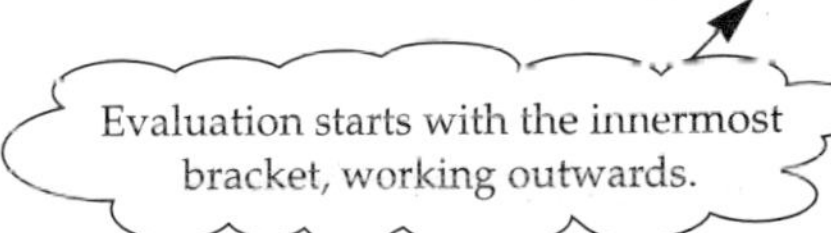

This involves just three multiplications and three additions compared with the six multiplications and three additions needed when evaluating $P(3)$ term by term. In Exercise 1A, question 17, we develop this further to produce a very compact and efficient way of evaluating a polynomial $P(x)$ at $x = a$; this not only tells us the remainder when $P(x)$ is divided by $x - a$, but also tells us the coefficients of the quotient.

Exercise 1A

1. Divide $x^3 + 4x^2 - 7x + 2$ by $x + 3$ and state the identity which is established.

2. Divide $x^4 + 6x^2 + 12$ by $x - 5$ and state the identity which is established.

3. Find the remainder when $2x^3 - 5x^2 - 4x + 9$ is divided by (i) $x + 4$; (ii) $2x - 1$.

4. Show that the remainder when polynomial $P(x)$ is divided by $ax + b$ is $P\left(-\frac{b}{a}\right)$.

5. The polynomial $P(x) \equiv x^3 + ax^2 + bx + c$ leaves remainders $-36, -20, 0$ on division by $x + 1, x + 2, x + 3$ respectively. Solve the equation $P(x) = 0$.

Exercise 1A continued

6. Divide $P(x)$ by $f(x)$ and express your result in the form of an identity.

(i) $P(x) \equiv 2x^4 - 3x^3 + 5x^2 - 5x - 3;$ $f(x) \equiv x^2 + 2$

(ii) $P(x) \equiv 4x^5 - 2x^4 - 2x^3 + x^2 - 3x + 2;$ $f(x) \equiv 2x^2 - 3$

(iii) $P(x) \equiv x^4 + 2x^3 - 5x^2 + 4x + 9;$ $f(x) \equiv x^2 - x + 3$

(iv) $P(x) \equiv 3x^4 - 8x^3 + 29x^2 + 21;$ $f(x) \equiv x^2 - x + 7$

7. (i) Write down an expression for the remainder when a polynomial $P(x)$ is divided by $(x - a)$.

When $f(x) = 2x^6 + kx^5 + 32x^2 - 26$ is divided by $(x + 1)$, the remainder is 15.

(ii) Calculate the value of the constant k.

$f(x)$ is now divided by $(x - 2)$, giving the quotient $g(x)$ and remainder R, so that

$$f(x) = (x - 2)g(x) + R. \qquad ①$$

(iii) Calculate the remainder R.

(iv) Calculate $g(-1)$.

(v) By first differentiating the identity ①, or otherwise, calculate $g(2)$.

[MEI]

8. What are the quotient and remainder when $x^4 + x + 1$ is divided by $x^2 + 1$?

9. If $(a + b + c)x^2 + (b - c)x + (c - 2) \equiv 0$ find the values of a, b, c.

10. (i) (a) Factorise $x^2 - a^2$ and $x^3 - a^3$.

(b) Show that $x^4 - a^4 \equiv (x - a)(x^3 + ax^2 + a^2x + a^3)$

(ii) By summing a suitable geometric progression prove that

$$x^n - a^n \equiv (x - a)(x^{n-1} + ax^{n-2} + a^2x^{n-3} + \ldots + a^{n-1}).$$

11. (i) (a) Show that $x + a$ is a factor of $x^3 + a^3$.

(b) Factorise $x^3 + a^3$ and $x^5 + a^5$.

(ii) Prove that provided n is odd

$$x^n + a^n \equiv (x + a)(x^{n-1} - ax^{n-2} + a^2x^{n-3} - \ldots - a^{n-2}x + a^{n-1}).$$

12. Show that one of the factors of $(x - y)^3$ is also a factor of $(y - z)^3 + (z - x)^3$ and so factorise $(x - y)^3 + (y - z)^3 + (z - x)^3$.

13. Explain the difference between the identity $P(x) \equiv 0$ and the equation $P(x) = 0$ when $P(x) \equiv ax^2 + bx + c$.

14. When the polynomial $P(x)$ is divided by $(x - a)(x - b)$ the quotient is $Q(x)$ and the remainder is $rx + s$. By writing this as an identity and giving suitable values to x find the constants r and s in terms of a, b, $P(a)$ and $P(b)$ assuming $a \neq b$.

15. When the polynomial $P(x)$ is divided by $(x - a)(x - b)$ the quotient is $Q(x)$ and there is no remainder. $P(x) = 0$ has no roots between a and b and the factors $(x - a)$, $(x - b)$ are not repeated. Prove that $Q(a)$ and $Q(b)$ have the same sign.

16. When $P_1(x)$ is divided by $x - a$ the remainder is R. When $P_2(x)$ is divided by $x - a$ the remainder is S.

(i) Prove that when $P_1(x) + P_2(x)$ is divided by $x - a$ the remainder is $R + S$.

(ii) Find the remainder when $P_1(x)P_2(x)$ is divided by $x - a$.

17. Horner's method of evaluating a polynomial at $x = a$ is as follows: write the coefficients of $P(x)$ in a row in descending order of the powers of x, including any coefficients which are 0; multiply the leading coefficient by a, and add the second coefficient; multiply this by a and add the next coefficient; repeat this process until you have reached the end of the row and added the constant coefficient. The final sum is the value of $P(a)$. The last numbers in the other columns are the respective coefficients of the quotient when $P(x)$ is divided by $x - a$.

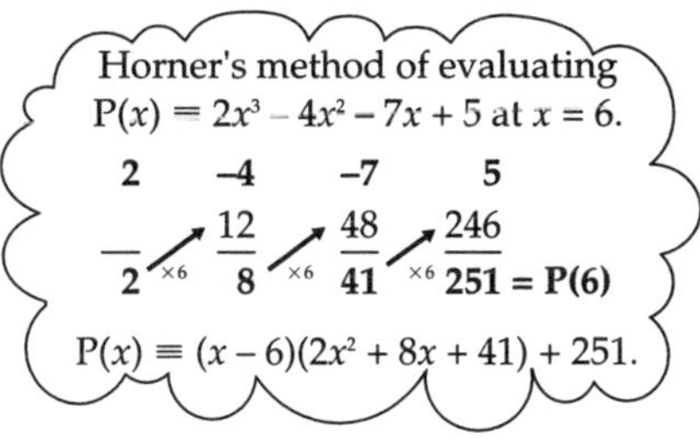

(i) Try the method out on several examples of your own choice.

(ii) Explain why it works.

(iii) Adapt the method to divide $P(x)$ by $ax - b$: you should first divide both $P(x)$ and $ax - b$ by a.

(Introduced by William George Horner in 1819, this method of dividing a polynomial by a linear factor is also known as *synthetic division*. It is worth noting that this process is easy to program on a spreadsheet.)

Extending the Factor Theorem

Suppose you want to find a cubic polynomial $P(x)$ which has three distinct roots: 1, 2, and 3. Then by the Factor Theorem $(x - 1)$ is a factor of $P(x)$, and so are $(x - 2)$ and $(x - 3)$.

$P(x)$ could be $(x - 1)(x - 2)(x - 3)$

or $5(x - 1)(x - 2)(x - 3)$

or $-2(x - 1)(x - 2)(x - 3)$

as illustrated in figure 1.2.

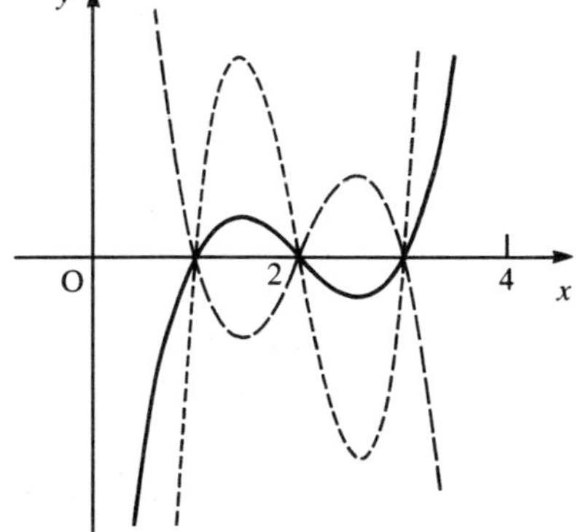

Figure 1.2

Although there are infinitely many possibilities for $P(x)$, you cannot find a cubic polynomial with roots 1, 2, and 3 which cannot be expressed in the form $P(x) \equiv c(x - 1)(x - 2)(x - 3)$, where c is a non-zero constant.

Extending and generalising this idea we now formally prove the following theorem by induction.

Theorem

If $\text{P}(x)$ is a polynomial of degree n ($\geqslant 1$) with n distinct roots $a_1, a_2, a_3, ..., a_n$ then $\text{P}(x) \equiv c(x - a_1)(x - a_2)(x - a_3)...(x - a_n)$, where c is a non-zero constant.

Proof

(i) We first prove that the theorem is true for $n = 1$.

$\text{P}(x)$ is a polynomial of degree $1 \Rightarrow \text{P}(x) \equiv c_1x + c_0$ where c_1 is non-zero.

$\text{P}(x)$ has root $a_1 \Rightarrow c_1a_1 + c_0 = 0 \Rightarrow c_0 = -c_1a_1$

so that $\text{P}(x) \equiv c_1x - c_1a_1 \equiv c_1(x - a_1)$ where c_1 is a non-zero constant, as required.

(ii) We now show that if the theorem is true for $n = k$ then it is also true for $n = k + 1$.

$\text{P}(x)$ is a polynomial of degree $k + 1$ with $k + 1$ distinct roots

$$a_1, a_2, a_3, ..., a_k, a_{k+1}$$

$\Rightarrow \text{P}(a_{k+1}) = 0$

$\Rightarrow \text{P}(x) \equiv (x - a_{k+1})\text{Q}(x)$

applying the Remainder Theorem

where $\text{Q}(x)$ is a polynomial of degree k.

Putting $x = a_1$ we have $(a_1 - a_{k+1})\text{Q}(a_1) = \text{P}(a_1) = 0$;

but $a_1 - a_{k+1} \neq 0$, so $\text{Q}(a_1) = 0$. In other words a_1 is a root of $\text{Q}(x)$.

In the same way we can show that $a_2, a_3, ..., a_k$ are also roots of $\text{Q}(x)$.

We have just shown that $\text{Q}(x)$ is of degree k, with k distinct roots $a_1, a_2, a_3, ... , a_k$ so we may apply the theorem:

$\text{Q}(x) \equiv c(x - a_1)(x - a_2)(x - a_3)...(x - a_k)$, where c is a non-zero constant.

Therefore $\text{P}(x) \equiv (x - a_{k+1})\text{Q}(x) \equiv (x - a_{k+1})c(x - a_1)(x - a_2)(x - a_3)...(x - a_k)$

$\equiv c(x - a_1)(x - a_2)(x - a_3)...(x - a_k)(x - a_{k+1})$

where c is a non-zero constant.

From (i) and (ii) by induction the theorem holds for all positive integers n.

One immediate consequence of this theorem is that a polynomial of degree n cannot have more than n distinct roots. For if $\text{P}(x)$ is a polynomial with n distinct roots $a_1, a_2, a_3, ..., a_n$, then $\text{P}(x) \equiv c(x - a_1)(x - a_2)(x - a_3)...(x - a_n)$, where c is a non-zero constant and $\text{P}(b) = c(b - a_1)(b - a_2)(b - a_3)...(b - a_n) \neq 0$ if b is not one of $a_1, a_2, a_3, ..., a_n$.

Equating coefficients

For Discussion

(i) Is it possible for the graphs of two different quadratic polynomials, $\text{P}(x)$ and $\text{Q}(x)$, to intersect in three points?

(ii) Is it possible for the graphs of two different cubic polynomials, $\text{P}(x)$ and $\text{Q}(x)$, to intersect in four points?

You have probably decided that if the graphs of $y = x^2$ and $y = ax^2 + bx + c$ intersect in three points then $a = 1$, and $b = c = 0$. In other words, the two graphs are not distinct.

Generalising this result: if P(x) and Q(x) are polynomials of the same degree, n, such that $P(x) = Q(x)$ for more than n distinct values of x then $P(x) \equiv Q(x)$. To prove this let $D(x) \equiv P(x) - Q(x)$. Then D(x) is a polynomial such that $D(x) = 0$ for more than n distinct values of x. That is: D(x) has more than n distinct roots which implies that D(x) cannot be a polynomial of degree n or less. The only possibility left is that $D(x) \equiv 0$

$$\Rightarrow P(x) - Q(x) \equiv 0$$

$$\Rightarrow P(x) \equiv Q(x).$$

Notice that as $P(x) - Q(x) \equiv 0$ all the coefficients of $P(x) - Q(x)$ are 0 so the coefficients of P(x) equal the corresponding coefficients of Q(x), giving rise to the process is known as *equating coefficients*.

For example: if $ax^3 + bx^2 - 5 \equiv p + qx + 6x^2 - 7x^3$

then we may equate coefficients, obtaining $a = -7$, $b = 6$, $p = -5$, $q = 0$.

We have said nothing about whether the numbers we are using are rational, real or complex. This is because the results we have obtained are general: they do not depend on the type of number in use. However when we deal with particular polynomials it can be important to know whether there are any restrictions on the numbers permitted.

Thus working with rational numbers:	$x^2 - 2$ has no factors;
working with real numbers:	$x^2 - 2 \equiv (x - \sqrt{2})(x + \sqrt{2})$.
Similarly, with real numbers:	$x^2 - 6x + 13$ has no factors;
with complex numbers:	$x^2 - 6x + 13 \equiv (x - 3 - 2j)(x - 3 + 2j)$.

EXAMPLE

Find, if possible, constants p, q, r

(i) such that $n^2 \equiv pn(n + 2) + q(n + 1) + r(n + 2)$;

(ii) such that $n^2 \equiv pn(n + 2) + q(n + 1)^2 + r$.

Solution

(i) Equating coefficients:

$$n^2: p = 1$$
$$n^1: 2p + q + r = 0$$
$$n^0: q + 2r = 0$$

These equations have the unique solution: $p = 1$, $q = -4$, $r = 2$.

(ii) Equating coefficients:

$$n^2: p + q = 1$$
$$n^1: 2p + 2q = 0$$
$$n^0: q + r = 0$$

The first two equations are inconsistent, so no such p, q, r can be found.

EXAMPLE

Prove that if a, b, c are distinct, then

$$P(x) \equiv \frac{(x-a)(x-b)}{(c-a)(c-b)} + \frac{(x-b)(x-c)}{(a-b)(a-c)} + \frac{(x-c)(x-a)}{(b-c)(b-a)} - 1 \equiv 0.$$

Solution

$$P(a) = 0 + \frac{(a-b)(a-c)}{(a-b)(a-c)} + 0 - 1 = 0.$$

Similarly $P(b) = P(c) = 0$.

$P(x)$ vanishes for three distinct values of x

$\Rightarrow P(x)$ cannot be a polynomial of degree less than 3.

By inspection $P(x)$ is of degree no more than 2.

Therefore degree of $P(x)$ is undefined. i.e. $P(x) \equiv 0$.

Exercise 1B

1. Find constants a, b, c so that $x^2 \equiv a(x-2)^2 + b(x-2) + c$.

2. Find constants a, b, c, d so that $x^3 \equiv a(x+1)^3 + b(x+1)^2 + c(x+1) + d$.

3. Show that there are no values of constants a, b such that

$$x^2 \equiv ax(x+1) + b(x+1)(x+2).$$

4. What can you say about the constants a, b, c in the following cases?

(i) $x \equiv a(x-1) + b(x-2) + c(x-3)$

(ii) $x^2 \equiv a(x-1)^2 + b(x-2)^2 + c(x-3)^2$

(iii) $x^3 \equiv a(x-1)^3 + b(x-2)^3 + c(x-3)^3$

5. Find constants a, b, c such that $(ar+1)^3 - (br-1)^3 \equiv 24r^2 + c$.

Hence show that $\sum_{r=1}^{n} r^2 \equiv \frac{1}{6}n(n+1)(2n+1)$.

6. (i) Show that the equation of the straight line through (a, A), (b, B) can be written in the form $y = \frac{A(x-b)}{a-b} + \frac{B(x-a)}{b-a}$, provided $a \neq b$.

(ii) Show that the equation of the one and only quadratic curve through (a, A), (b, B), (c, C) is

$$y = \frac{A(x-b)(x-c)}{(a-b)(a-c)} + \frac{B(x-c)(x-a)}{(b-c)(b-a)} + \frac{C(x-a)(x-b)}{(c-a)(c-b)}$$

where a, b, c are distinct.

(iii) Write down the equation of the unique cubic curve which goes through (a, A), (b, B), (c, C), (d, D), where a, b, c, d are distinct.

(This method is due to Joseph-Louis Lagrange, 1736–1813.)

7. Use the results of Question 6 to find

(i) the equation of the quadratic curve through $(1, 1)$, $(2, 5)$, $(3, 15)$;

(ii) the cubic polynomial $P(x)$ such that $P(-1) = -5$, $P(1) = 1$, $P(2) = 1$, $P(3) = 7$.

8. A quadratic approximation for 2^x is required in the interval $1 \leqslant x \leqslant 5$.

By considering $\mathrm{h}(x) = A(x-3)(x-5) + B(x-1)(x-5) + C(x-1)(x-3)$, where A, B, and C are constants, or otherwise, find the quadratic function $\mathrm{h}(x)$ such that $\mathrm{h}(x) = 2^x$ when $x = 1, 3$ and 5.

Give your answer in the form $\mathrm{h}(x) = ax^2 + bx + c$. [MEI]

9. Prove that $\dfrac{a(x-b)(x-c)}{(a-b)(a-c)} + \dfrac{b(x-c)(x-a)}{(b-c)(b-a)} + \dfrac{c(x-a)(x-b)}{(c-a)(c-b)} \equiv x$ provided a, b, c are distinct.

10. Prove that $\dfrac{a^2-x^2}{(a-b)(a-c)} + \dfrac{b^2-x^2}{(b-c)(b-a)} + \dfrac{c^2-x^2}{(c-a)(c-b)} \equiv 1$ provided a, b, c are distinct.

11. Prove that $\dfrac{a^2(x-b)(x-c)}{(a-b)(a-c)} + \dfrac{b^2(x-c)(x-a)}{(b-c)(b-a)} + \dfrac{c^2(x-a)(x-b)}{(c-a)(c-b)} \equiv x^2$ provided a, b, c are distinct.

12. (i) Prove that: m has an odd factor (other than 1) $\Rightarrow 2^m + 1$ is not prime.

[**Hint:** let $m = ab$, where b is an odd number greater than 1, so that $2^m + 1 = (2^a)^b + 1$ and then use the identity $x^n + y^n \equiv (x+y)(x^{n-1} - x^{n-2}y + x^{n-3}y^2 - \ldots - xy^{n-2} + y^{n-1})$, provided n is odd.]

(ii) Deduce that $2^m + 1$ is prime $\Rightarrow m = 2^n$ for some $n \geqslant 0$.

(Numbers of the form $2^{2^n} + 1$ are known as Fermat numbers.)

13. Integers greater than 1 which are not prime are described as *composite*.

(i) Prove that: a is an integer greater than $2 \Rightarrow a^m - 1$ is composite.

(ii) Prove that: m is composite $\Rightarrow 2^m - 1$ is composite.

(iii) (a) Prove that: $a^p - 1$ is prime $\Rightarrow p$ is prime and $a = 2$.

(b) Let $M_p = 2^p - 1$. Show that M_2, M_3, M_5, M_7 are prime but that M_{11} is composite.

(Primes of the form $2^p - 1$ are called Mersenne primes, after Marin Mersenne, 1588–1648. At the time of writing, the largest known prime is $M_{1\,257\,787}$, with 378 632 digits, discovered in September 1996.)

14. (i) Show that the value of $x^2 + 5x + 5$ is prime for $x = 0, 1, 2, 3, 4$.

(ii) All the coefficients of the quadratic $\mathrm{P}(x)$ are integers.

(a) Suppose $\mathrm{P}(x) \equiv (px+q)(rx+s)$ where p, q, r, s are integers.

Explain why: a is an integer such that $\mathrm{P}(a)$ is a prime number $\Rightarrow$ either $pa + q = \pm 1$ or $ra + s = \pm 1$.

(b) Show that: $\mathrm{P}(x)$ is prime for five distinct integer values of x $\Rightarrow \mathrm{P}(x)$ cannot be factorised over the set of integers.

(iii) $\mathrm{S}(x)$ is a polynomial of degree n with integer coefficients. Find a similar sufficient condition for it to be impossible to factorise $\mathrm{S}(x)$ into linear factors over the integers.

Repeated roots

For Discussion

The diagram shows the graph of $y = f(x)$.

What can you say about the roots of the two equations

$$f(x) = 0$$

$$f'(x) = 0?$$

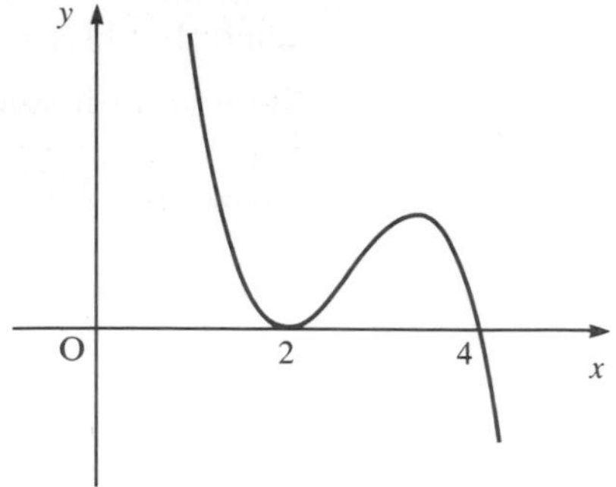

Figure 1.3

Having a little extra information often eases the problem when solving an equation, as illustrated in the next example.

EXAMPLE

Solve the equation $x^3 + x^2 - 33x + 63 = 0$ given that it has repeated roots.

Solution

Let $f(x) \equiv x^3 + x^2 - 33x + 63$.

Then $f'(x) \equiv 3x^2 + 2x - 33 \equiv (3x + 11)(x - 3)$.

Since $f(x) = 0$ has a repeated root, $f(x) = 0$ and $f'(x) = 0$ have a common root.

The roots of $f'(x) = 0$ are $x = 3$ and $x = -\frac{11}{3}$;

$f(3) = 3^3 + 3^2 - 33 \times 3 + 63 = 27 + 9 - 99 + 63 = 0$ so $x = 3$ is root of $f(x) = 0$.

As $f(x)$ is a cubic only one of the roots of $f'(x) = 0$ can also satisfy $f(x) = 0$, so we do not need to evaluate $f\left(-\frac{11}{3}\right)$.

Now $x^3 + x^2 - 33x + 63 = 0$

$\Leftrightarrow (x - 3)^2(x + 7) = 0$

so that roots are $x = 3$ (repeated) and $x = -7$.

> Since $x = 3$ is a repeated root, $(x - 3)$ is a repeated factor. Observe that $\frac{63}{(-3)^2} = 7$ gives second term in final factor.

Exercise 1C

1 – 7. These equations have repeated roots. Use that fact to help you solve them.

1. $x^3 - 10x^2 + 32x - 32 = 0$

2. $x^3 - 15x^2 + 72x - 108 = 0$

3. $2x^3 + 11x^2 + 12x - 9 = 0$

4. $9x^3 + 39x^2 - 29x + 5 = 0$

5. $4x^3 - 27x - 27 = 0$

6. $x^4 + x^3 - 3x^2 - 5x - 2 = 0$

7. $x^4 + 6x^3 + 13x^2 + 12x + 4 = 0$

8. Use calculus to prove that:

$$f(x) \equiv (x - a)^2 g(x) \Rightarrow f'(x) \text{ has a factor } (x - a).$$

What happens if $f(x) \equiv (x - a)^n g(x)$?

Properties of the roots of polynomial equations

Quadratic equations

In the work which follows we are using z as the variable (or unknown) instead of x to emphasise that our results apply regardless of whether the roots are complex or real. Suppose that α and β are the roots of the quadratic equation $az^2 + bz + c = 0$, $a \neq 0$. Then by the theorem we proved on page 10

$$az^2 + bz + c \equiv a(z - \alpha)(z - \beta) \equiv az^2 - a(\alpha + \beta)z + a\alpha\beta.$$

Equating coefficients shows that

$$\text{the sum of the roots} = \alpha + \beta = -\frac{b}{a}$$

$$\text{and the product of the roots} = \alpha\beta = \frac{c}{a}.$$

From these results we can obtain information about the roots without actually solving the equation.

EXAMPLE

The roots of the equation $2z^2 + 3z + 8 = 0$ are α and β. Find

(i) the sum and product of α and β;

(ii) the quadratic equation with roots α^2 and β^2.

Solution

(i) $\alpha + \beta = -\frac{3}{2}$. ← sum of roots $= -\frac{b}{a}$

$\alpha\beta = \frac{8}{2} = 4$. ← product of roots $= \frac{c}{a}$

(ii) *Method 1*

$$(\alpha + \beta)^2 = \alpha^2 + \beta^2 + 2\alpha\beta \Rightarrow (-\tfrac{3}{2})^2 = \alpha^2 + \beta^2 + 2 \times 4$$

$$\Rightarrow \alpha^2 + \beta^2 = \tfrac{9}{4} - 8 = -\tfrac{23}{4}.$$

$\alpha^2\beta^2 = (\alpha\beta)^2 = 16.$

The required equation is $z^2 - (\alpha^2 + \beta^2)z + \alpha^2\beta^2 = 0$

i.e. $z^2 - (-\frac{23}{4})z + 16 = 0$ ← sum of roots; product of roots

which is perhaps better written as $4z^2 + 23z + 64 = 0$.

Method 2: Here is an alternative way of finding $\alpha^2 + \beta^2$.

α is root of $2z^2 + 3z + 8 = 0 \Rightarrow 2\alpha^2 + 3\alpha + 8 = 0$.

β is root of $2z^2 + 3z + 8 = 0 \Rightarrow 2\beta^2 + 3\beta + 8 = 0$.

Adding: $2(\alpha^2 + \beta^2) + 3(\alpha + \beta) + 16 = 0$

$$\Rightarrow 2(\alpha^2 + \beta^2) = -3(\alpha + \beta) - 16 = \tfrac{9}{2} - 16 = -\tfrac{23}{2}$$

$\Rightarrow \alpha^2 + \beta^2 = -\frac{23}{4}$, and then proceed as above.

Activity

What happens if you try to solve the quadratic equation $az^2 + bz + c = 0$ by solving the equations $\alpha + \beta = -\frac{b}{a}$, $\alpha\beta = \frac{c}{a}$ simultaneously?

Exercise 1D

1. Write down the sum and product of the roots of each quadratic equation:

(i) $2z^2 + 7z + 6 = 0$; (ii) $5z^2 - z - 1 = 0$;

(iii) $7z^2 + 2 = 0$; (iv) $5z^2 + 24z = 0$;

(v) $z(z + 8) = 4 - 3z$; (vi) $3z^2 + 8z - 6 = 0$.

2. Write down quadratic equations (with integer coefficients) with roots

(i) $7, 3$; (ii) $-5, -4.5$;

(iii) $5, 0$; (iv) 3 repeated;

(v) $3 - 2j, 3 + 2j$.

3. The roots of $z^2 - 2z + 3 = 0$ are α and β.

(i) Write down the values of $\alpha + \beta$ and $\alpha\beta$ and deduce the values of $\frac{1}{\alpha} + \frac{1}{\beta}$ and $\frac{1}{\alpha} \times \frac{1}{\beta}$.

(ii) Write down the equation (with integer coefficients) whose roots are $\frac{1}{\alpha}, \frac{1}{\beta}$.

(iii) Find the equation whose roots are the reciprocals of the roots of $az^2 + bz + c = 0$.

4. The roots of $2z^2 + 5z - 9 = 0$ are α and β. Find quadratic equations with roots

(i) $-\alpha$ and $-\beta$; (ii) $2\alpha + \beta$ and $\alpha + 2\beta$;

(iii) α^2 and β^2; (iv) $\frac{\alpha}{\beta}$ and $\frac{\beta}{\alpha}$.

5. The roots of the equation $z^2 + 8z - 2 = 0$ are α and β. Find the equation with roots $\alpha^2\beta$ and $\alpha\beta^2$.

6. The roots of $az^2 + bz + c = 0$ are α and β. Find quadratic equations with roots

(i) $k\alpha$ and $k\beta$;

(ii) $k + \alpha$ and $k + \beta$.

7. Using the fact that $\alpha + \beta = -\frac{b}{a}$, $\alpha\beta = \frac{c}{a}$, what can you say about the roots α and β of $az^2 + bz + c = 0$ if you also know that

(i) a, b, c are all positive and $b^2 - 4ac > 0$?

(ii) $b = 0$?

(iii) $c = 0$?

(iv) a and c have opposite signs?

8. One root of $az^2 + bz + c = 0$ is twice the other. Prove that $2b^2 = 9ac$.

9. (i) The straight line $y = 2x + k$ meets the parabola $y = 3x^2 - 4x - 11$ at the points P and Q. Form a quadratic equation for the x co-ordinates of P and Q. Without solving this equation find the x co-ordinate of M, the midpoint of PQ. What do you deduce about the locus of M as k varies?

(ii) By considering the locus of the midpoint of chords of fixed gradient m, generalise this result to the parabola $y = ax^2 + bx + c$.

Cubic equations

There are corresponding properties for the roots of cubic and quartic equations (as well as equations of higher degree). If α, β, γ are the roots of the cubic equation $az^3 + bz^2 + cz + d = 0$ then, as before,

$$az^3 + bz^2 + cz + d \equiv a(z - \alpha)(z - \beta)(z - \gamma)$$
$$\equiv az^3 - a(\alpha + \beta + \gamma)z^2 + a(\alpha\beta + \beta\gamma + \gamma\alpha)z - a\alpha\beta\gamma.$$

Equating coefficients we obtain these results for roots of cubic equations:

$$\alpha + \beta + \gamma = -\frac{b}{a},$$ sum of individual roots

$$\alpha\beta + \beta\gamma + \gamma\alpha = \frac{c}{a},$$ sum of products of roots in pairs

$$\alpha\beta\gamma = -\frac{d}{a}.$$ product of roots

As with the roots of quadratic equations, we cannot find the roots directly from these equations as attempting to solve them simultaneously merely leads us back to the original cubic equation (with α or β or γ in place of z). But if we have additional information these equations can provide a quick and easy method of solution.

NOTE

For brevity we often use $\Sigma\alpha$ and $\Sigma\alpha\beta$ to denote $\alpha + \beta + \gamma$ and $\alpha\beta + \beta\gamma + \gamma\alpha$ respectively. There is no ambiguity provided we know the degree of the relevant equation. Functions like these are called symmetric functions of the roots, because interchange of any two of α, β, γ leaves their value unchanged. Similar notation is used to denote other symmetric functions of the roots. For example: the sum of all products of one root with the square of another

$$= \Sigma\alpha^2\beta = \alpha^2\beta + \alpha\beta^2 + \alpha^2\gamma + \alpha\gamma^2 + \beta^2\gamma + \beta\gamma^2.$$

Activity

Prove the following identities for cubics:

(i) $(\Sigma\alpha)^2 \equiv \Sigma\alpha^2 + 2\Sigma\alpha\beta$;

(ii) $\alpha\beta\gamma\Sigma\alpha^{-1} \equiv \Sigma\alpha\beta$;

(iii) $\Sigma\alpha^3 - 3\alpha\beta\gamma \equiv (\Sigma\alpha)(\Sigma\alpha^2 - \Sigma\alpha\beta)$.

EXAMPLE

Solve the equation $2z^3 - 9z^2 - 27z + 54 = 0$ given that the roots form a geometric progression.

Solution

Let the three roots be $\frac{\alpha}{r}$, α, αr.

You could use $\alpha, \alpha r, \alpha r^2$ but our choice makes for simpler equations as the product of our three roots does not contain r.

The product of the roots $= \alpha^3 = -\frac{54}{2} = -27 \Leftrightarrow \alpha = -3$.

$$\text{The sum of the roots } = \Sigma\alpha = \frac{\alpha}{r} + \alpha + \alpha r = \tfrac{9}{2}$$

$$\Leftrightarrow 2r^2 + 5r + 2 = 0$$

$$\Leftrightarrow (2r+1)(r+2) = 0$$

$$\Leftrightarrow r = -2 \text{ or } r = -\tfrac{1}{2}.$$

Substituting for α, multiplying by $2r/3$ and rearranging.

Both values of r tell us that the three roots are $\frac{3}{2}$, -3, 6.

EXAMPLE

The roots of the cubic equation $az^3 + bz^2 + cz + d = 0$ are α, β, γ, where $\alpha\beta\gamma \neq 0$. Find the cubic equation with roots $\frac{1}{\alpha}, \frac{1}{\beta}, \frac{1}{\gamma}$.

Solution

Method 1

$$\frac{1}{\alpha} + \frac{1}{\beta} + \frac{1}{\gamma} = \frac{\beta\gamma + \gamma\alpha + \alpha\beta}{\alpha\beta\gamma} = \frac{c/a}{-d/a} = -\frac{c}{d}.$$

$$\frac{1}{\alpha}\times\frac{1}{\beta} + \frac{1}{\beta}\times\frac{1}{\gamma} + \frac{1}{\gamma}\times\frac{1}{\alpha} = \frac{\gamma + \alpha + \beta}{\alpha\beta\gamma} = \frac{-b/a}{-d/a} = \frac{b}{d}.$$

$$\frac{1}{\alpha}\times\frac{1}{\beta}\times\frac{1}{\gamma} = \frac{1}{\alpha\beta\gamma} = -\frac{a}{d}.$$

So the required equation is

$$z^3 - \left(-\frac{c}{d}\right)z^2 + \frac{b}{d}z - \left(-\frac{a}{d}\right) = 0 \Leftrightarrow dz^3 + cz^2 + bz + a = 0.$$

Method 2 (by substitution)

Let $w = \frac{1}{z}$ so that $z = \frac{1}{w}$. Then

α, β, γ are the roots of $az^3 + bz^2 + cz + d = 0$ if and only if

$\frac{1}{\alpha}, \frac{1}{\beta}, \frac{1}{\gamma}$ are the roots of $a \times \frac{1}{w^3} + b \times \frac{1}{w^2} + c \times \frac{1}{w} + d = 0$

$$\Leftrightarrow a + bw + cw^2 + dw^3 = 0.$$

Notice that taking the reciprocal of the roots reverses the coefficients of the equation. This property applies to polynomial equations of all degrees, provided that none of the roots is zero.

The substitution method used above is powerful. We illustrate it again in the next example.

EXAMPLE

The roots of the cubic equation $az^3 + bz^2 + cz + d = 0$ are α, β, γ. Find the cubic equation with roots (i) $3\alpha + 7, 3\beta + 7, 3\gamma + 7$; (ii) $\alpha^2, \beta^2, \gamma^2$.

Solution

(i) Let $w = 3z + 7$ so that $z = \dfrac{w-7}{3}$. Then α, β, γ are the roots of

$az^3 + bz^2 + cz + d = 0$ if and only if $3\alpha + 7, 3\beta + 7, 3\gamma + 7$

are the roots of $a\left(\dfrac{w-7}{3}\right)^3 + b\left(\dfrac{w-7}{3}\right)^2 + c\left(\dfrac{w-7}{3}\right) + d = 0$

$\Leftrightarrow a(w^3 - 21w^2 + 147w - 343) + 3b(w^2 - 14w + 49) + 9c(w - 7) + 27d = 0$

$- 343a - 147b + 63c$ which simplifies to

$$aw^3 - 3(7a - b)w^2 + 3(49a - 14b + 3c)w - (343a - 147b + 63c - 27d) = 0.$$

(ii) Let $w = z^2$. Then

$az^3 + bz^2 + cz + d = 0 \Rightarrow awz + bw + cz + d = 0$

Substituting w for z^2 wherever it occurs.

$\Rightarrow (aw + c)z = -(bw + d)$

Again substituting w for z^2.

$\Rightarrow (aw + c)^2w = (bw + d)^2$

$\Rightarrow a^2w^3 + (2ac - b^2)w^2 + (c^2 - 2bd)w - d^2 = 0.$

Activity

The cubic equation $az^3 + bz^2 + cz + d = 0$ has roots α, β, γ. Explain why substituting $-\dfrac{d}{aw}$ for z in $az^3 + bz^2 + cz + d = 0$ forms an equation with roots $\alpha\beta, \beta\gamma, \gamma\alpha$. Simplify the resulting equation as much as possible and show that your result is valid even if one of α, β, γ is 0.

Our next example shows how we may, on occasions, convert a set of three simultaneous equations (in three unknowns) into a single cubic equation. Finding a single root of the cubic (perhaps by trial and improvement) is often easier than finding a set of three values that satisfy the original equations simultaneously.

EXAMPLE

Solve the simultaneous equations

$x + y + z = 2; \qquad x^2 + y^2 + z^2 = 62; \qquad x^3 + y^3 + z^3 = 92.$

Solution

Suppose x, y, z are the roots α, β, γ of the cubic equation $t^3 + bt^2 + ct + d = 0$.

$\Sigma\alpha = x + y + z = 2 \Rightarrow b = -2.$

See the Activity on page 17.

$(\Sigma\alpha)^2 \equiv \Sigma\alpha^2 + 2\Sigma\alpha\beta$

$\Rightarrow \Sigma\alpha\beta = ((\Sigma\alpha)^2 - \Sigma\alpha^2)/2 = (2^2 - 62)/2 = -29$ so that $c = -29$.

$\Sigma\alpha^3 - 3\alpha\beta\gamma \equiv (\Sigma\alpha)(\Sigma\alpha^2 - \Sigma\alpha\beta)$

See page 17.

$\Rightarrow 3\alpha\beta\gamma = \Sigma\alpha^3 - (\Sigma\alpha)(\Sigma\alpha^2 - \Sigma\alpha\beta) = 92 - 2(62 + 29) = -90$

$\Rightarrow \alpha\beta\gamma = -30$ so that $d = 30$.

We now need to solve $t^3 - 2t^2 - 29t + 30 = 0$.

By inspection one root is $t = 1$, so that $t - 1$ is a factor: $(t - 1)(t^2 - t - 30) = 0$

$$\Leftrightarrow (t - 1)(t - 6)(t + 5) = 0 \Leftrightarrow t = 1, 6 \text{ or } -5.$$

Thus $x = 1$, $y = 6$, $z = -5$ or any equivalent permutation.

Quartic equations

We can adapt our method of treating the roots of quadratic and cubic equations to quartic equations. If $\alpha, \beta, \gamma, \delta$ are the roots of

$az^4 + bz^3 + cz^2 + dz + e = 0$, where $a \neq 0$, then

$$az^4 + bz^3 + cz^2 + dz + e \equiv a(z - \alpha)(z - \beta)(z - \gamma)(z - \delta)$$

$$\equiv az^4 - a(\alpha + \beta + \gamma + \delta)z^3 + a(\alpha\beta + \alpha\gamma + \alpha\delta + \beta\gamma + \beta\delta + \gamma\delta)z^2$$

$$- a(\alpha\beta\gamma + \beta\gamma\delta + \gamma\delta\alpha + \delta\alpha\beta)z + a\alpha\beta\gamma\delta.$$

Equating coefficients shows that:

$$\Sigma\alpha = \alpha + \beta + \gamma + \delta = -\frac{b}{a},$$ sum of individual roots

$$\Sigma\alpha\beta = \alpha\beta + \alpha\gamma + \alpha\delta + \beta\gamma + \beta\delta + \gamma\delta = \frac{c}{a},$$ sum of products of roots in pairs

$$\Sigma\alpha\beta\gamma = \alpha\beta\gamma + \beta\gamma\delta + \gamma\delta\alpha + \delta\alpha\beta = -\frac{d}{a},$$ sum of products of roots in threes

$$\alpha\beta\gamma\delta = \frac{e}{a}.$$ product of roots

EXAMPLE

The roots of the quartic equation $z^4 + 3z^3 - 2z^2 - z + 5 = 0$ are $\alpha, \beta, \gamma, \delta$. Find $\Sigma\alpha^3$.

Solution

$\alpha\beta\gamma\delta = 5 \Rightarrow$ none of $\alpha, \beta, \gamma, \delta$ is zero.

$\Sigma\alpha = -\frac{b}{a} = -3.$

$\Sigma\alpha\beta = \frac{c}{a} = -2.$

α is a root $\Rightarrow \alpha^4 + 3\alpha^3 - 2\alpha^2 - \alpha + 5 = 0 \Rightarrow \alpha^3 + 3\alpha^2 - 2\alpha - 1 + 5\alpha^{-1} = 0$

$$\Rightarrow \alpha^3 = -3\alpha^2 + 2\alpha + 1 - 5\alpha^{-1}$$

with similar expressions for $\beta^3, \gamma^3, \delta^3$. Adding these four expressions gives

$$\Sigma\alpha^3 = -3\Sigma\alpha^2 + 2\Sigma\alpha + 4 - 5\Sigma\alpha^{-1}$$

We are adding four expressions.

$$= -3[(\Sigma\alpha)^2 - 2\Sigma\alpha\beta] + 2\Sigma\alpha + 4 - 5\Sigma\alpha^{-1}$$

$\Sigma\alpha^2 \equiv (\Sigma\alpha)^2 - 2\Sigma\alpha\beta$

$$= -3[(-3)^2 - 2(-2)] + 2(-3) + 4 - 5 \times \tfrac{1}{5}$$

$$= -42.$$

$\Sigma\alpha^{-1} = -\frac{d}{e} = \frac{1}{5}$ since taking the reciprocal of the roots reverses the coefficients of the equation.

Exercise 1E

1. The roots of $2z^3 + 3z^2 - z + 7 = 0$ are α, β, γ. Find

(i) $\Sigma\alpha$;
(ii) $\Sigma\alpha\beta$;
(iii) $\alpha\beta\gamma$;
(iv) $\Sigma\alpha^2$;
(v) $\Sigma\alpha^3$;
(vi) $\Sigma\alpha^4$;
(vii) $\Sigma\frac{1}{\alpha}$;
(viii) $\Sigma\frac{1}{\alpha\beta}$;
(ix) $\Sigma\alpha^2\beta$;
(x) $\Sigma\frac{\alpha+\beta}{\gamma}$.

2. The roots of $z^3 - 4z^2 - z + 3 = 0$ are α, β, γ. Find cubic equations whose roots are

(i) $2\alpha, 2\beta, 2\gamma$;

(ii) $\alpha + 2, \beta + 2, \gamma + 2$;

(iii) $\alpha + \beta, \beta + \gamma, \gamma + \alpha$.

3. Solve these equations given that the roots are in arithmetic progression.

(i) $z^3 - 15z^2 + 66z - 80 = 0$

(ii) $9z^3 - 18z^2 - 4z + 8 = 0$

(iii) $z^3 - 6z^2 + 16 = 0$

(iv) $54z^3 - 189z^2 + 207z - 70 = 0$

4. The roots of the equation $2z^3 - 12z^2 + kz - 15 = 0$ are in arithmetic progression. Solve the equation and find k.

5. Solve $32z^3 - 14z + 3 = 0$, given that one root is twice another.

6. The roots of $az^3 + bz^2 + cz + d = 0$ are α, β, γ. Form cubic equations with roots

(i) $\alpha^2, \beta^2, \gamma^2$;
(ii) $\frac{1}{\alpha^2}, \frac{1}{\beta^2}, \frac{1}{\gamma^2}$;
(iii) $\frac{\beta\gamma}{\alpha}, \frac{\gamma\alpha}{\beta}, \frac{\alpha\beta}{\gamma}$.

7. Find a formula connecting a, b, c, d which is a necessary and sufficient condition for the roots of the equation $az^3 + bz^2 + cz + d = 0$ to be in geometric progression. Show that this condition is satisfied for the equation $8z^3 - 52z^2 + 78z - 27 = 0$, and hence solve the equation.

8. The roots of the cubic equation $x^3 - 5x^2 - 6x - 4 = 0$ are α, β and γ.

(i) Write down the values of $\alpha + \beta + \gamma$, $\alpha\beta + \beta\gamma + \gamma\alpha$ and $\alpha\beta\gamma$.

(ii) Find the value of $\alpha^2 + \beta^2 + \gamma^2$.

(iii) Show that $(\alpha\beta)^2 + (\beta\gamma)^2 + (\gamma\alpha)^2 = -4$. Deduce that α, β and γ are not all real.

(iv) Find a cubic equation with integer coefficients whose roots are $\frac{\beta\gamma}{\alpha}, \frac{\gamma\alpha}{\beta}$ and $\frac{\alpha\beta}{\gamma}$.

[MEI]

Algebra

9. The equation $z^3 + pz^2 + 2pz + q = 0$ has roots $\alpha, 2\alpha, 4\alpha$. Find all the possible values of p, q, α.

10. Solve the following simultaneous equations.

(i)
$$x + y + z = -1$$
$$xy + yz + zx = -17$$
$$xyz = -15$$

(ii)
$$x + y + z = 6$$
$$xy + yz + zx = 5$$
$$xyz = -12$$

(iii)
$$x + y + z = 2$$
$$xy + yz + zx = -9$$
$$xyz = -18$$

11. Solve the following simultaneous equations.

(i)
$$x + y + z = 2$$
$$x^2 + y^2 + z^2 = 30$$
$$x^3 + y^3 + z^3 = 116$$

(ii)
$$x + y + z = 2$$
$$x^2 + y^2 + z^2 = 34$$
$$x^3 + y^3 + z^3 = 98$$

(iii)
$$x + y + z = 5$$
$$x^2 + y^2 + z^2 = 7$$
$$x^3 + y^3 + z^3 = 5$$

12. Show that: one root of $az^3 + bz^2 + cz + d = 0$ is the reciprocal of another root if and only if $a^2 - d^2 = ac - bd$.

Verify that this condition is satisfied for the equation $21z^3 - 16z^2 - 95z + 42 = 0$ and hence solve the equation.

13. The roots of $z^3 + pz^2 + qz + r = 0$ are $\alpha, -\alpha, \beta$, and $r \neq 0$. Show that $r = pq$, and find all three roots in terms of p and q.

14. The roots of $2z^4 + 3z^3 + 6z^2 - 5z + 4 = 0$ are $\alpha, \beta, \gamma, \delta$. Find

(i) $\Sigma\alpha^2$; (ii) $\Sigma(\alpha + \beta + \gamma)^2$;

(iii) $\Sigma\alpha\beta^2$; (iv) $\Sigma(\alpha + \beta)^3$.

15. Prove the following identities for quartics.

(i) $(\Sigma\alpha)^2 \equiv \Sigma\alpha^2 + 2\Sigma\alpha\beta$

(ii) $\alpha\beta\gamma\delta\,\Sigma\alpha^{-1} \equiv \Sigma\alpha\beta\gamma$

(iii) $\Sigma\alpha^3 - 3\Sigma\alpha\beta\gamma \equiv (\Sigma\alpha)(\Sigma\alpha^2 - \Sigma\alpha\beta)$

16. The roots of $z^4 + pz^3 + qz^2 + rz + s = 0$ are $\alpha, \beta, \gamma, \delta$. Prove that: $\alpha\beta = \gamma\delta \Rightarrow s = r^2/p^2$.

17. The roots of $az^4 + bz^3 + cz^2 + dz + e = 0$ are $\alpha, \beta, \gamma, \delta$. Show that

$$a\,\Sigma\alpha^{n+4} + b\,\Sigma\alpha^{n+3} + c\,\Sigma\alpha^{n+2} + d\,\Sigma\alpha^{n+1} + e\,\Sigma\alpha^n = 0.$$

Investigations

1. When the non-zero coefficients of the polynomial $\mathrm{P}(x)$ are arranged in descending order of the corresponding powers of x, there are s places where the coefficients change sign. Investigate *Descartes' rule of signs* which states that the number of positive roots of the equation $\mathrm{P}(x) = 0$ is either s or $s - 2k$, where k is a positive integer. Find a similar way of deciding an upper limit for the number of negative roots of $\mathrm{P}(x) = 0$.

2. Investigate ways of extending Horner's method of synthetic division to dividing a polynomial by a quadratic. (See Exercise 1A, question 17.)

3. Investigate the following iterative procedure for solving the quadratic equation $ax^2 + bx + c = 0$: first choose a starting value (α_0); then use $\alpha_n + \beta_n = -\frac{b}{a}$ and $\alpha_{n+1}\beta_n = \frac{c}{a}$ to find β_0 and other values of α_n and β_n.

K E Y P O I N T S

- The Remainder Theorem:

  If polynomial $\mathrm{P}(x)$ is divided by $x - a$ the remainder is $\mathrm{P}(a)$.

- A polynomial of degree n cannot take the value zero for more than n distinct values of x.

- If $\mathrm{P}(x)$ and $\mathrm{Q}(x)$ are polynomials of the same degree, n, such that $\mathrm{P}(x) = \mathrm{Q}(x)$ for more than n distinct values of x then $\mathrm{P}(x) \equiv \mathrm{Q}(x)$.

- If α and β are the roots of the quadratic equation $az^2 + bz + c = 0$, then

$$\alpha + \beta = -\frac{b}{a},$$

$$\alpha\beta = \frac{c}{a}.$$

- If α, β, γ are the roots of the cubic equation $az^3 + bz^2 + cz + d = 0$, then

$$\Sigma\alpha = \alpha + \beta + \gamma = -\frac{b}{a},$$

$$\Sigma\alpha\beta = \alpha\beta + \beta\gamma + \gamma\alpha = \frac{c}{a},$$

$$\alpha\beta\gamma = -\frac{d}{a}.$$

- If $\alpha, \beta, \gamma, \delta$ are the roots of the quartic equation $az^4 + bz^3 + cz^2 + dz + e = 0$, then

$$\Sigma\alpha = \alpha + \beta + \gamma + \delta = -\frac{b}{a},$$

$$\Sigma\alpha\beta = \alpha\beta + \alpha\gamma + \alpha\delta + \beta\gamma + \beta\delta + \gamma\delta = \frac{c}{a},$$

$$\Sigma\alpha\beta\gamma = \alpha\beta\gamma + \beta\gamma\delta + \gamma\delta\alpha + \delta\alpha\beta = -\frac{d}{a},$$

$$\alpha\beta\gamma\delta = \frac{e}{a}.$$

2 Geometry with polar co-ordinates

Μηδεὶς ἀγεωμέτρητος εἰσίτω μου τὴν στέγην
[let no one ignorant of geometry enter my door].

Inscription over the entrance to the Academy of Plato, c.430–349 BC

This Nautilus shell forms an equiangular spiral. How could you describe this mathematically?

Polar co-ordinates

When dealing with the polar form of a complex number (*Pure Mathematics 4*, page 73) you used the idea of describing the position of a point P in a plane by giving its distance r from a fixed point O and the angle θ between OP and a fixed direction. In this system, first used by Newton in 1671, O is called the *pole* and the angle θ is measured from the initial line which is usually drawn to the right across the page, like the positive x axis; the numbers (r, θ) are called the *polar co-ordinates* of P, figure 2.1.

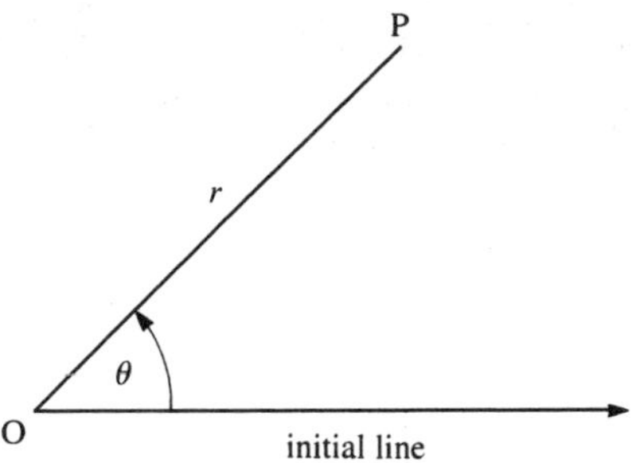

Figure 2.1

As before the angle θ is positive in the anticlockwise sense from the initial line; at the pole itself $r = 0$ and θ is undefined. Each pair of numbers (r, θ) gives a unique point, but the converse is not true, for two reasons. Firstly, a point is not changed if we add any integer multiple of 2π to the angle θ. Secondly, it is sometimes convenient to let r take negative values (something which does not happen with complex numbers since $|z| \geqslant 0$), with the natural interpretation that the point $(-r, \theta)$ is the same as $(r, \theta + \pi)$.

Activity

Check by drawing a diagram that the polar co-ordinates

$(5, \pi/3)$, $(5, 7\pi/3)$, $(5, -11\pi/3)$ and $(-5, -2\pi/3)$ all describe the same point.

Give three other pairs of polar co-ordinates for the point $(-6, 3\pi/4)$.

If it is necessary to specify the polar co-ordinates of a point uniquely then we use those for which $r > 0$ and $-\pi < \theta \leqslant \pi$; these are called the *principal polar co-ordinates*.

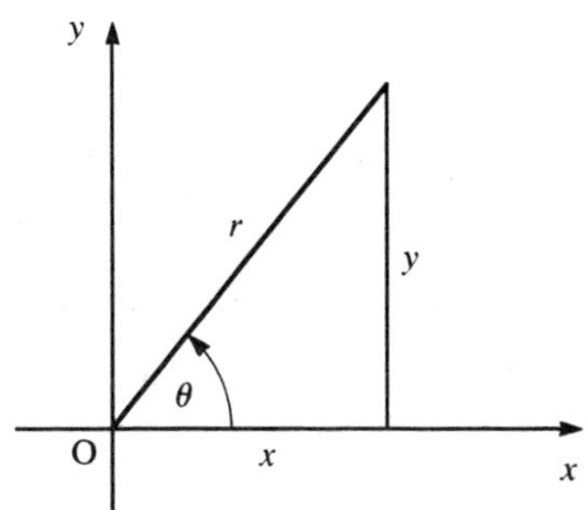

Figure 2.2

It is easy to change between polar co-ordinates (r, θ) and Cartesian co-ordinates (x, y) since, from figure 2.2,

$$x = r\cos\theta \qquad y = r\sin\theta$$

$$r = \sqrt{x^2 + y^2} \qquad \tan\theta = \frac{y}{x}$$

You need to be careful to choose the right quadrant when finding θ, since the equation $\tan\theta = \frac{y}{x}$ always gives two solutions, differing by π. Always draw a sketch to check which one of these is correct.

Exercise 2A

1. Plot the points A, B, C, D with polar co-ordinates $(3, \pi/5)$, $(2, 7\pi/10)$, $(3, -4\pi/5)$, $(-4, 7\pi/10)$ respectively. What shape is ABCD?

2. One vertex of an equilateral triangle has polar co-ordinates $(4, \pi/4)$. Find the polar co-ordinates of all the possible other vertices

(a) when the origin O is the centre of the triangle;

(b) when O is another vertex of the triangle;

(c) when O is the midpoint of one side of the triangle.

3. The diagram below shows a regular pentagon OABCD, in which A has Cartesian co-ordinates $(5, 2)$.

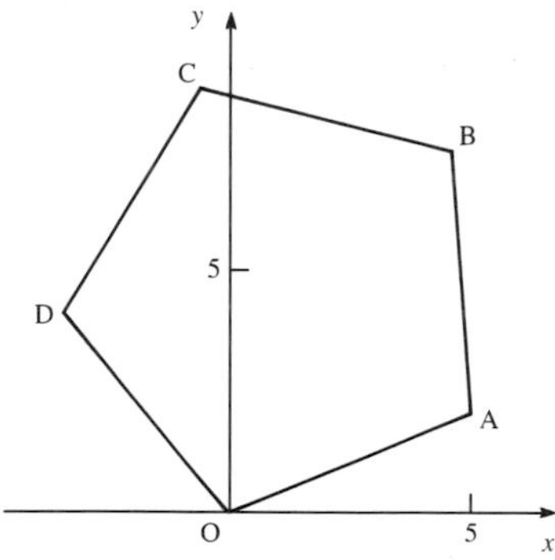

(i) Show that OB = 8.71 (correct to two decimal places).

(ii) Find the polar co-ordinates of A, B, C, D.

(iii) Hence find the Cartesian co-ordinates of B, C, D.

[In (ii) and (iii) give your answers correct to two decimal places.]

4. In this question r is in millimetres and θ is in degrees. The scoring region of a dartboard is marked by six concentric circles, called inner bull, outer bull, inner treble, outer treble, inner double, outer double, with radii 6, 16, 99, 107, 162, 170 mm respectively (to the nearest mm, ignoring the thickness of the dividing wire). The part between the outer bull and outer double circles is divided into twenty equal 'sectors', numbered as shown below, and the board is hung with the 20 sector vertically above the centre so that the initial line bisects the 6 sector. A dart scores 50 in the inner bull and 25 in the outer bull, where $6 < r < 16$. A dart in a sector scores the sector number, except that within the doubles ring $(162 < r < 170)$ or trebles ring $(99 < r < 107)$ it scores double or treble the sector number respectively.

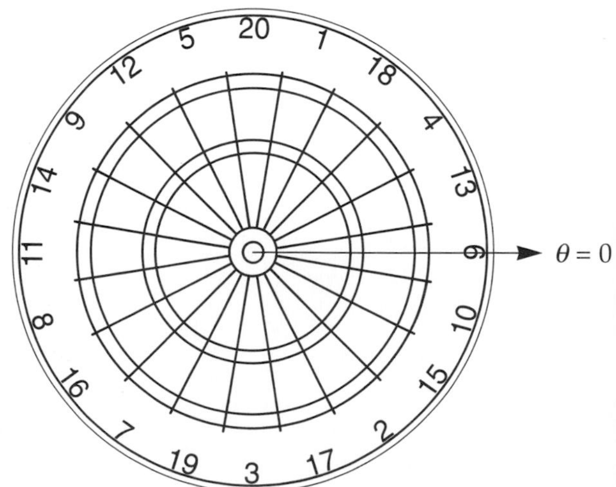

(i) Find the score in the region for which $16 < r < 99$ and $27 < \theta < 45$.

(ii) Give conditions on r and θ which define the boundary between sectors 10 and 15.

(iii) Give conditions on r and θ for the regions in which the score is

(a) treble 14; (b) 17; (c) 18.

The polar equation of a curve

The points (r, θ) for which the values of r and θ are linked by a function f form a curve whose *polar equation* is $r = \mathrm{f}(\theta)$. The polar equation of a curve may be simpler than its Cartesian equation, especially if the curve has rotational symmetry. Polar equations have many important applications, for example in the study of orbits (see *Mechanics 5*).

EXAMPLE

Investigate the curve with polar equation $r = 10\cos\theta$.

Solution

We tackle this in three ways.

(i) *By plotting*. Make a table of values. This one has θ increasing by $\pi/12$ (i.e. 15°), which gives a convenient number of points.

θ	0	$\pi/12$	$\pi/6$	$\pi/4$	$\pi/3$	$5\pi/12$	$\pi/2$	$7\pi/12$	$2\pi/3$	$3\pi/4$	$5\pi/6$	$11\pi/12$	π
r	10	9.7	8.7	7.1	5.0	2.6	0	–2.6	–5.0	–7.1	–8.7	–9.7	–10

Values of θ from 0 to $-\pi$ (or from π to 2π) give the same points again: for example, $\theta = -\pi/12 \Rightarrow r = 9.7$, which is the same point as $(-9.7, 11\pi/12)$. Plotting these points gives the curve shown below.

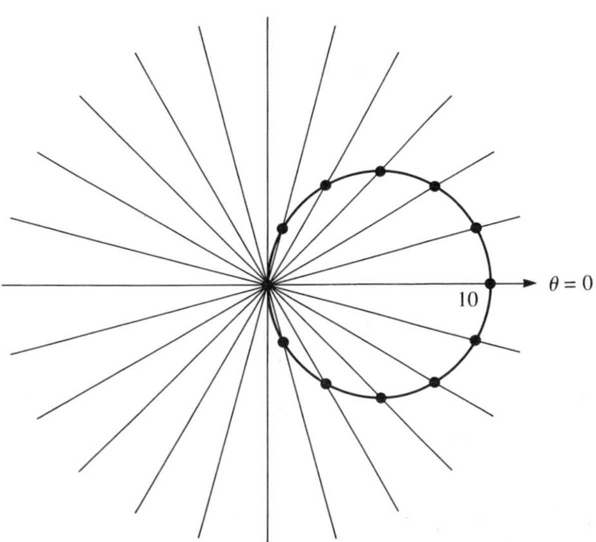

(ii) *By converting to Cartesian form.* If $r \neq 0$ then

$$r = 10\cos\theta \iff r^2 = 10r\cos\theta$$
$$\iff x^2 + y^2 = 10x.$$

If $r = 0$ then $x = y = 0$, which also satisfies $x^2 + y^2 = 10x$.

Therefore the Cartesian equation is $x^2 + y^2 = 10x$

$$\iff (x-5)^2 + y^2 = 25,$$

which shows that the curve is the circle with radius 5 and centre (5, 0) (in Cartesian co-ordinates).

(iii) *By geometrical reasoning.* Knowing the answer leads to an even simpler solution. If P is the point on this circle with polar co-ordinates (r, θ) then OPA is a right angle (the angle in a semicircle) and so $r = 10\cos\theta$ as required, and the same applies to points on the lower semicircle since the cosine function is an even function.

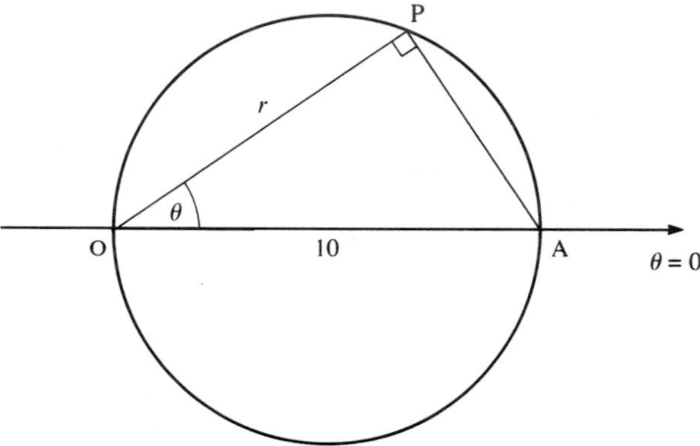

NOTES

1. *Plotting and joining points as in (i) above gives a good idea of the shape of the curve, but the argument in (ii) or (iii) is needed before you can be sure that this is truly a circle.*
2. *As the value of θ increases from $-\pi$ to π the point moves twice around the circle.*

Activity

If you have access to a graphics calculator or a computer with suitable software, find out how to draw a curve from its polar equation. Check that you can adjust the scales so that in this case you get a circle, not just an ellipse.

For Discussion

Some graphics calculators will not draw the curve $r = \mathrm{f}(\theta)$ directly, but instead you can take θ as a parameter and draw the curve with parametric equations $x = \mathrm{f}(\theta)\cos\theta,\ y = \mathrm{f}(\theta)\sin\theta$. Explain why this works.

EXAMPLE

(i) Describe the motion of a point along the curve $r = 1 + 2\cos\theta$ as θ increases from 0 to 2π.

(ii) Do the same for the curve $r = \dfrac{1}{1+2\cos\theta}$.

Solution

(i) The curve is shown in the diagram.

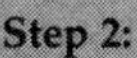

Step 2:

As θ increases to $2\pi/3$, r decreases to zero since $\cos(2\pi/3) = -\frac{1}{2}$.

Step 1:

As θ increases from 0 to $\pi/2$, r decreases from 3 to 1.

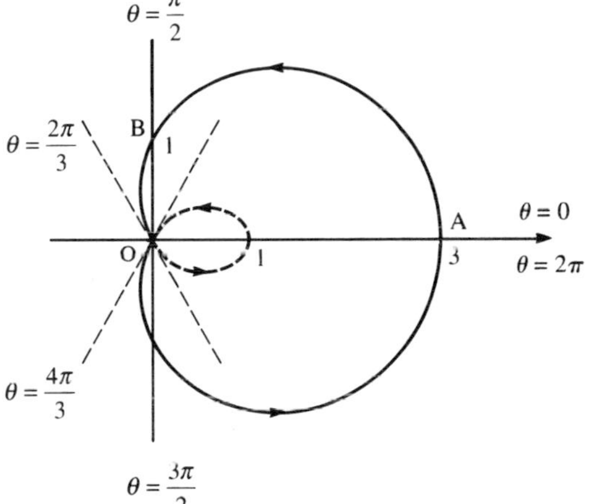

Step 3:

The curve touches the line $\theta = 2\pi/3$ at O.

Step 4:

For θ between $2\pi/3$ and $4\pi/3$, r is negative, with $r = -1$ at $\theta = \pi$.

Step 5:

The curve touches the line $\theta = 4\pi/3$ at O.

Step 6:

r increases to 1 at $\theta = 3\pi/2$ and then to 3 at $\theta = 2\pi$.

This double loop is one of a family of curves called *limaçons* (snail curves); see the investigation on page 36.

(ii) The value of r is now the reciprocal of the value found in (i); the curve is shown overleaf.

This curve has two separate branches; it is an example of a hyperbola, and will be dealt with in detail later in this chapter (page 33).

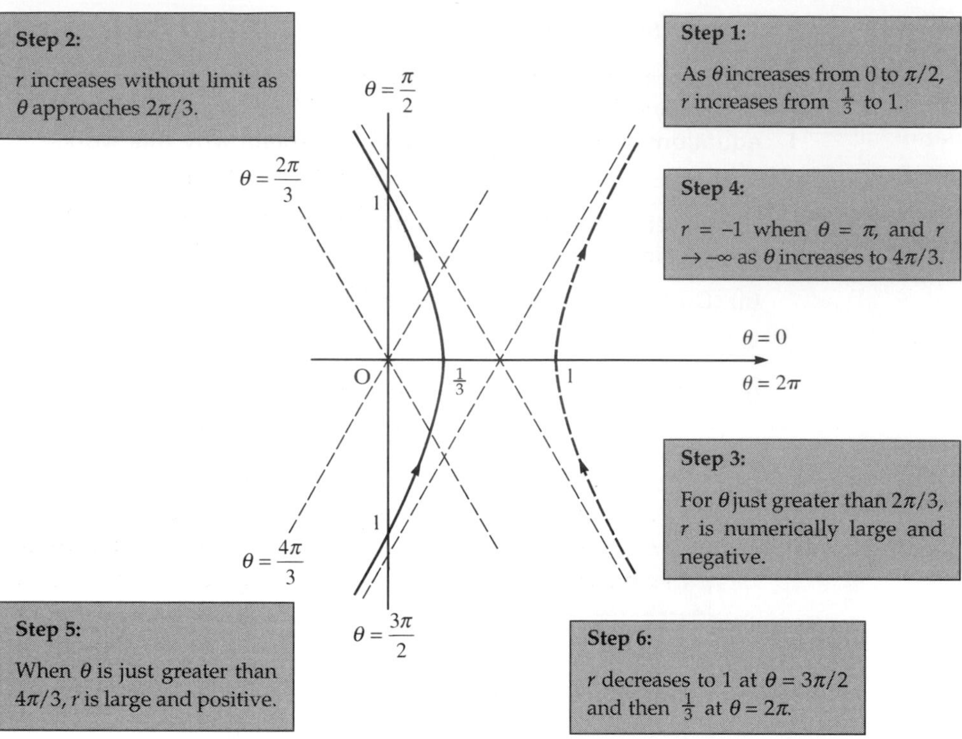

NOTE

The diagrams in the example above use the convention that the parts of the curve for which $r < 0$ are shown by a broken line. In some applications it is physically impossible for r to be negative, so it is worth distinguishing such portions in this way.

Exercise 2B

In this exercise you should make full but critical use of a graphics calculator or computer if these are available.

1. Make a table of values of $8\sin\theta$ for θ from 0 to π at intervals of $\pi/12$ $(= 15^\circ)$, and say what happens when $\pi \leqslant \theta \leqslant 2\pi$. By plotting points draw the curve $r = 8\sin\theta$.

Prove that this curve is a circle, and give its Cartesian equation.

2. Draw the graph of the *spiral of Archimedes*

$r = \dfrac{4\theta}{\pi}$ for $-2\pi \leqslant \theta \leqslant 2\pi$.

3. A curve with polar equation $r = k\sin n\theta$, where k and n are positive and n is an integer, is called a *rhodonea* (rose curve). Throughout this question take $k = 10$.

(i) What shape is the curve when $n = 1$?

(ii) Draw the curve when $n = 2$.

(iii) Draw the curve when $n = 3$.

(iv) From these examples (and others if you wish) form a conjecture about how the number of 'petals' depends on n.

4. A curve with polar equation $r = a(1 + \cos\theta)$ is called a *cardioid*. Draw the curve when $a = 8$, and account for its name.

5. Prove that $r = a\sec\theta$ and $r = b\,\text{cosec}\,\theta$, where a and b are non-zero constants, are the polar equations of two straight lines. Find their Cartesian equations.

Exercise 2B continued

6. The straight line ℓ passes through the point A with polar co-ordinates (p, α) and is perpendicular to OA. Prove that the polar equation of ℓ is $r\cos(\theta - \alpha) = p$.

Use the expansion of $\cos(\theta - \alpha)$ to find the Cartesian equation of ℓ.

7. Sketch on the same diagram the curves with polar equations $r = 2a\cos\theta$, $2r(1 + \cos\theta) = 3a$ and find the polar co-ordinates of their points of intersection.

What is the polar equation of the common chord of the two curves? [MEI]

The area of a sector

The region bounded by an arc UV of a curve and the two lines OU and OV is called a *sector*. In order to find the area of the sector for which OU and OV are the lines $\theta = \alpha$ and $\theta = \beta$ and the curve is $r = f(\theta)$ we first divide it into small sectors such as OPQ, where P and Q have co-ordinates (r, θ) and $(r + \delta r, \theta + \delta\theta)$, as in figure 2.4.

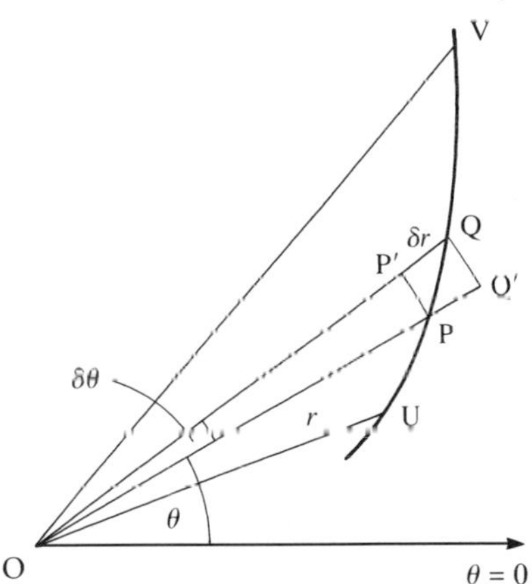

Figure 2.4

Let the areas of sectors OUP and OPQ be A and δA respectively. Then since the area of sector OPQ lies between the area of the circular sectors OPP′ and OQQ′,

$$\tfrac{1}{2}r^2\delta\theta < \delta A < \tfrac{1}{2}(r + \delta r)^2\delta\theta$$

Remember that θ is in radians.

and so $\tfrac{1}{2}r^2 < \dfrac{\delta A}{\delta\theta} < \tfrac{1}{2}(r + \delta r)^2$.

Now as $\delta\theta \to 0$, $\dfrac{\delta A}{\delta\theta} \to \dfrac{dA}{d\theta}$, the rate of change of A with respect to θ.

But $\dfrac{\delta A}{\delta\theta}$ is trapped between $\frac{1}{2}r^2$, which is fixed, and $\frac{1}{2}(r+\delta r)^2$, which tends to $\frac{1}{2}r^2$, and so $\dfrac{\delta A}{\delta\theta}$ must also tend to $\frac{1}{2}r^2$. Therefore

$$\frac{dA}{d\theta}=\tfrac{1}{2}r^2.$$

From this key result the area of the sector can be found by integration:

$$\text{area OUV}=\int_{\alpha}^{\beta}\tfrac{1}{2}r^2 d\theta.$$

For Discussion

The argument given above is based on figure 2.4 in which

(i) $\delta\theta$ is positive, (ii) r increases as θ increases.

Consider how the argument must be adapted if

(a) $\delta\theta$ is negative, (b) r decreases as θ increases, (c) both (a) and (b).

Note that the final result remains the same in all cases.

EXAMPLE

Find the area of the inner loop of the limaçon $r = 1 + 2\cos\theta$ drawn on page 29.

Solution

The inner loop is formed as θ varies from $2\pi/3$ to $4\pi/3$, so its area is

$$\int_{2\pi/3}^{4\pi/3}\tfrac{1}{2}(1+2\cos\theta)^2 d\theta = \int_{2\pi/3}^{4\pi/3}\tfrac{1}{2}(1+4\cos\theta+4\cos^2\theta)d\theta$$

$$=\int_{2\pi/3}^{4\pi/3}(\tfrac{1}{2}+2\cos\theta+(1+\cos 2\theta))d\theta$$

using $\cos^2\theta=\frac{1}{2}(1+\cos 2\theta)$

$$=\left[\frac{3\theta}{2}+2\sin\theta+\tfrac{1}{2}\sin 2\theta\right]_{2\pi/3}^{4\pi/3}$$

$$=\pi-\frac{3\sqrt{3}}{2}.$$

NOTE

Even though r is negative for $2\pi/3 < \theta < 4\pi/3$, the integrand $\frac{1}{2}r^2$ is always positive, so there is no problem of 'negative areas' as there is with curves below the x axis in Cartesian co-ordinates.

Activity

For the limaçon $r = 1 + 2\cos\theta$ find

(i) the total area contained by the outer loop

(ii) the area between the two loops.

Exercise 2C

1. Check that $\int \frac{1}{2}r^2 \mathrm{d}\theta$ gives the area of the circle $r = 10\cos\theta$ correctly when the integral is evaluated from $-\pi/2$ to $\pi/2$ or from 0 to π. What happens when the integration is from 0 to 2π?

2. Find the area bounded by the spiral $r = \dfrac{4\theta}{\pi}$ from $\theta = 0$ to $\theta = 2\pi$ and the initial line.

3. Find the areas of the two portions into which the line $\theta = \pi/2$ divides the upper half of the cardioid $r = 8(1 + \cos\theta)$.

4. The diagram below shows the *equiangular spiral* $r = a\mathrm{e}^{k\theta}$, where a and k are positive constants, and the lines $\theta = 0$ and $\theta = \pi/4$. Prove that the areas of the regions $A, B, C, \ldots$ between these lines and successive whorls form a geometric sequence, and find its common ratio.

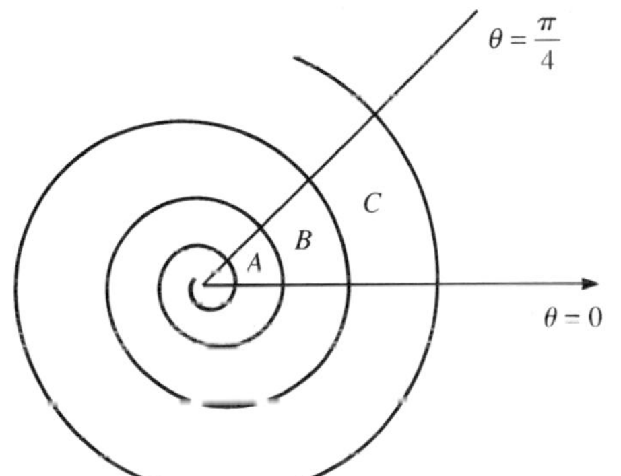

5. Sketch *Bernoulli's lemniscate* (ribbon bow) $r^2 = a^2\cos 2\theta$, and find the area of one of its loops.

6. The interior of the circle $r = 3a\cos\theta$ is divided into two parts by the cardioid $r = a(1 + \cos\theta)$. Find the area of the part whose boundary passes through the origin. [MEI]

7. A curve is defined by the parametric equations $x = \mathrm{f}(t)$, $y = \mathrm{g}(t)$. By differentiating the relation $\tan\theta = \dfrac{y}{x}$ with respect to t show that $r^2\dfrac{\mathrm{d}\theta}{\mathrm{d}t} = x\dfrac{\mathrm{d}y}{\mathrm{d}t} - y\dfrac{\mathrm{d}x}{\mathrm{d}t}$.
As t increases from t_1 to t_2 the point on the curve moves from P_1 to P_2, and θ increases. Prove that the area of the sector $\mathrm{OP}_1\mathrm{P}_2$ is

$$\frac{1}{2}\int_{t_1}^{t_2}\left(x\frac{\mathrm{d}y}{\mathrm{d}t} - y\frac{\mathrm{d}x}{\mathrm{d}t}\right)\mathrm{d}t.$$

8. The arc PQ is defined by $x = t^2$, $y = t^3$, $1 \leqslant t \leqslant 2$. Use question 7 to find the area of the sector bounded by this arc, OP and OQ.

9. Sketch the *astroid* $x = a\cos^3 t$, $y = a\sin^3 t$, and find the area it encloses.

10. Prove that the area enclosed by the curve $x = a\cos t + b\sin t$, $y = c\cos t + d\sin t$ is $\pi|ad - bc|$.

Conics

The family of curves called *conics* occupies a central place in mathematics, having a long history, a rich geometry, and many important applications. All this springs from a simple definition:

> a *conic* is the locus of a point in a plane such that its distance from a fixed point S is a constant multiple of its distance from a fixed line d, both S and d being in the plane.

The fixed point is called the *focus*, the fixed line is the *directrix*, and the constant multiplier is the *eccentricity*, denoted by e. The reasons for these names will appear in Chapter 5. If P is a point of the conic and M is the foot of the perpendicular from P to d then this focus–directrix definition says that SP = ePM.

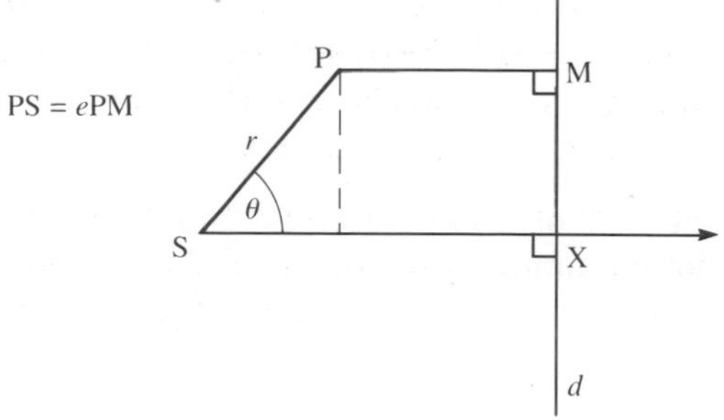

Figure 2.5

This leads immediately to the polar equation of a conic. Take the focus S as the pole, and the perpendicular SX from S to the directrix d as the initial line (figure 2.5). Then

$$
\begin{aligned}
\text{P lies on the conic} \quad &\Leftrightarrow \quad \text{SP} = e\text{PM} \\
&\Leftrightarrow \quad r = e(\text{SX} - r\cos\theta) \\
&\Leftrightarrow \quad r(1 + \cos\theta) = e\text{SX} \\
&\Leftrightarrow \quad \frac{\ell}{r} = 1 + e\cos\theta, \text{ where } \ell \text{ is the constant } e\text{SX}.
\end{aligned}
$$

The constant ℓ has a simple geometrical meaning: when $\theta = \pi/2$, $\cos\theta = 0$ and so $r = \ell$.

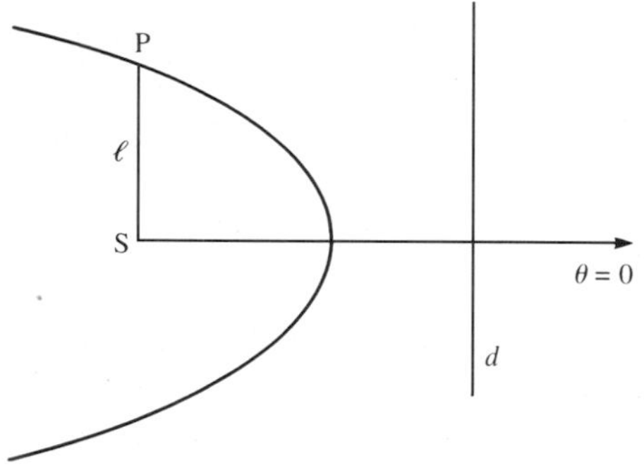

Figure 2.6

The chord of the conic through the focus and parallel to the directrix is called the *latus rectum* ('upright side'), so ℓ is the length of the semi-latus rectum (figure 2.6)

The polar equation can be rearranged to give

$$r = \frac{\ell}{1 + e\cos\theta},$$

which in turn shows there are three distinct types of conic, depending on the size of e.

1. If $0 < e < 1$ then the denominator $1 + e\cos\theta$ is never zero, so r remains finite, giving a closed curve which is called an *ellipse*.
2. If $e = 1$ then the denominator is zero when $\theta = \pi$, so that r can be made arbitrarily large as θ approaches π. The conic is a curve open towards the $\theta = \pi$ direction, called a *parabola*.

3. If $e > 1$ then $1 + e\cos\theta = 0$ when $\cos\theta = -1/e$. This gives two values of θ for which r is undefined, and a range of values for which r is negative. The conic has two branches, and is called a *hyperbola*. The example on page 29 shows one hyperbola in detail.

These three types of conic are shown in figure 2.7; it is clear from the definition that they are all symmetrical about the initial line, but we have no reason so far for assuming any further symmetry.

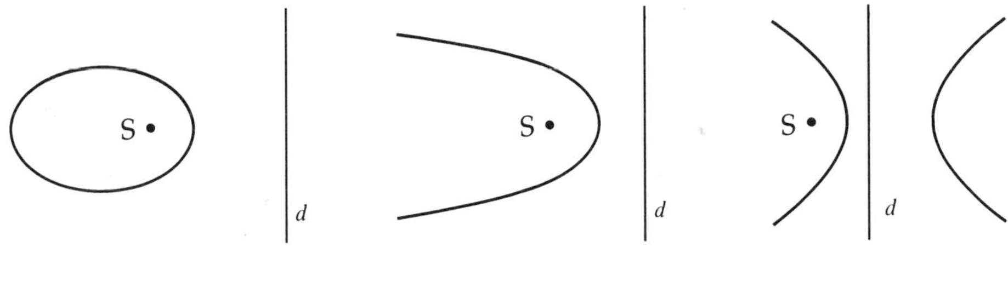

(1) Ellipse, $e < 1$ (2) Parabola, $e = 1$ (3) Hyperbola, $e > 1$

Figure 2.7

NOTE

You have already met these conics in Pure Mathematics 1 and Pure Mathematics 3, where they seem to have little connection with what we have just done. This subject has many different starting points, and one of the challenges (which will be taken up in Chapter 5) is to show that the various possible definitions of conics all lead to the same curves.

HISTORICAL NOTE

The Greek mathematician Apollonius of Perga (c.262–190 BC) wrote an eight volume study of conics, building on earlier work. Pappus of Alexandria first emphasised the focus–directrix properties in about AD320. The astronomer Johannes Kepler gave conics new importance when he announced in 1609 that the orbits of the planets are ellipses with the sun at the focus; the conventional use of the letter S (initial of solus = sun) to denote the focus is due to him.

Exercise 2D

Throughout this exercise C is the conic with polar equation $\frac{\ell}{r} = 1 + e\cos\theta$.

1. By plotting points or using a graphics calculator or computer, draw separate sketch graphs of C when $\ell = 5$ and

   (i) $e = \frac{1}{3}$; (ii) $e = \frac{1}{2}$,

   (iii) $e = 1$; (iv) $e = 2$;

   (v) $e = 3$.

2. Prove that the distance from the focus to the directrix is ℓ/e and add the directrix to each of your sketches in question 1. Show that the polar equation of the directrix can be put in the form $\frac{\ell}{r} = e\cos\theta$.

3. A family of conics all have a common focus and directrix. Sketch on a single diagram the members of this family for which

   (i) $e = \frac{1}{3}$; (ii) $e = \frac{1}{2}$;

   (iii) $e = 1$; (iv) $e = 2$;

   (v) $e = 3$.

4. Describe C (i) when $e = 0$; (ii) when e is very large.

Exercise 2D continued

5. In polar co-ordinates (r, θ) two conics C_1 and C_2 have equations

$$\frac{a}{r} = 1 + \cos\theta \quad \text{and} \quad \frac{4a}{r} = 3 + 2\cos\theta.$$

(i) Find the polar co-ordinates of the two points where C_1 and C_2 intersect.

(ii) Sketch C_1 and C_2 on the same diagram, giving a clear indication of the scale.

[MEI, part]

6. Prove that the polar equation of the parabola can be written in the form

$$r = \frac{\ell}{2}\sec^2\frac{\theta}{2}.$$

7. A chord PQ of C passes through the focus S.

Prove that $\dfrac{1}{\text{PS}} + \dfrac{1}{\text{QS}} = \dfrac{2}{\ell}$.

A chord of a conic which passes through a focus is called a *focal chord.*

8. The chords PQ, UV of a parabola or an ellipse are perpendicular focal chords.

Prove that $\dfrac{1}{\text{PQ}} + \dfrac{1}{\text{UV}} = \dfrac{2-e^2}{2\ell}$.

9. (i) Prove that $\dfrac{\ell}{r} = \cos(\theta - \alpha) + e\cos\theta$ is the polar equation of a straight line.

[**Hint:** convert to Cartesian co-ordinates.]

(ii) Show that this line meets $\boldsymbol{C}$ where $\theta = \alpha$ and nowhere else.

(iii) Deduce that this line is the tangent to $\boldsymbol{C}$ at the point where $\theta = \alpha$.

10. Prove that the tangents at the end of a focal chord meet on the directrix.

11. The tangents from a point T to $\boldsymbol{C}$ touch $\boldsymbol{C}$ at H and K.

Prove that angle TSH = angle TSK, where S is the focus.

Investigation

Limaçons

(i) Figure 2.8 shows a circle with centre A and diameter OB = $2a$. A line through O meets the circle again at Q, and P, P′ are points on this line such that PQ = QP′ = k (a fixed distance). Draw this figure, taking $a = 3$ cm and $k = 8$ cm. Draw many positions of the line as Q moves around the circle, and draw a freehand curve through the marked points P and P′. This curve is called a *limaçon.* Draw another limaçon with $a = k = 3$ cm.

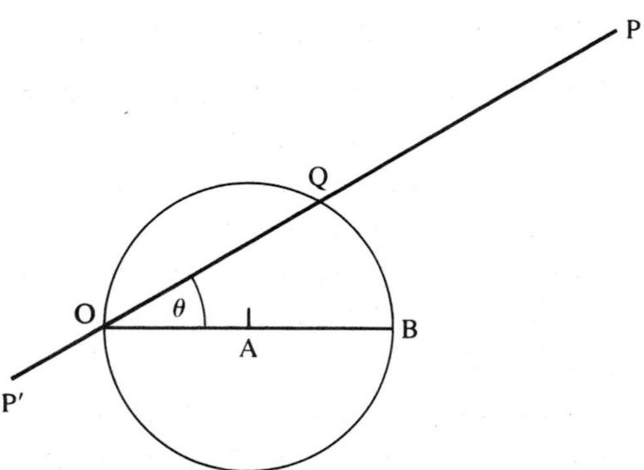

Figure 2.8

(ii) Without further detailed drawing, sketch other limaçons, noting the general shape (a) when $k > 2a$, (b) when $k < 2a$. What happens when k is very large, or when k is close to zero?

(iii) Prove that the polar equation of the limaçon is $r = k + 2a\cos\theta$, and explain how this gives the point P′ as well as the point P.

(iv) Find the area enclosed by the limaçon $r = k + 2a\cos\theta$ when $k \geqslant 2a$.

(v) Prove that the special limaçon for which $k = 2a$ is a cardioid.

(vi) The special limaçon for which $k = a$ is called the *trisectrix*; this is the curve drawn in part (i) of the example on page 29. As its name indicates, the trisectrix can be used for trisecting an angle (one of the famous Greek construction problems, which cannot be solved by straight edge and compasses alone), as follows.

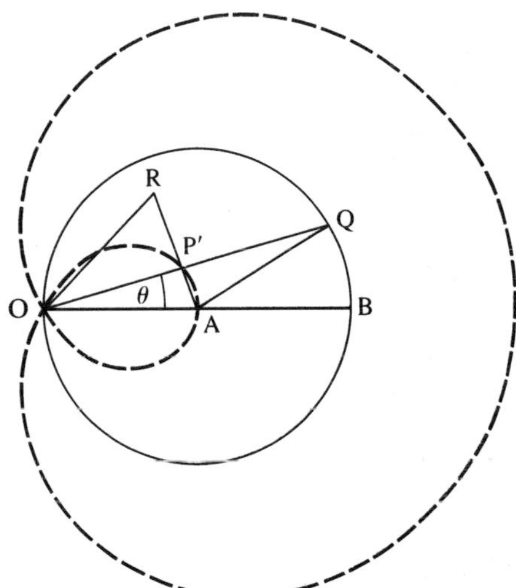

Figure 2.9

Draw angle BOR equal to the angle which is to be trisected, making OR equal to the radius a. Let RA meet the inner loop of the trisectrix at P′. Then angle BOP′ is one third of angle BOR.

Prove this by finding angles OQA, OAQ, QAP′, RAO, BOR in terms of θ.

[**Hint:** look for three isosceles triangles.]

(vii) The limaçon is also the locus of a point fixed at a distance a from the centre of a circular disc of radius $k/2$ which rolls without slipping around a fixed circle also of radius $k/2$. To prove this, copy figure 2.9 and complete the parallelogram PQAA′. Then draw the circle C with centre A and radius $k/2$, and the circle C' with centre A′ and radius $k/2$. Let these circle touch at T, let C meet AB at V, and let PA′ produced meet C' at V′.

Prove that arc TV = arc TV′, so that C' can be rolled around C until V′ coincides with V. Thus V′ is a point fixed on the rolling circle C'. But P is a fixed distance a from A′ along V′A′ since A′P = AQ, and therefore P is a fixed point of the circular disc bounded by C'. Note that if $k = 2a$ then P is on the circle C' (giving a cardioid), and that if $k < 2a$ then P is outside C', as if on the flange of a wheel.

(viii) Yet another way of drawing a limaçon is to let O be a fixed point at a distance a from the centre of a base-circle of radius $k/2$. Then with any point T on the base-circle as centre, draw the circle which passes through O. These circles all touch the limaçon.

Draw a limaçon by this method – it is worth doing carefully, for the result is beautiful. If you want to see why this works, look again at the diagram in (vii), and explain why TO = TP and why P moves at right angles to TP.

HISTORICAL NOTE

These curves first appear in a book published by Albrecht Dürer in 1525, drawn by what is essentially the rolling circle method of (vii). Between 1630 and 1640 G.P. de Roberval developed pre-calculus methods for drawing tangents: he named one of his examples the 'limaçon de monsieur Pascal', referring to its inventor Étienne Pascal, the father of Blaise Pascal.

KEY POINTS

- The principal polar co-ordinates (r, θ) are those for which $r > 0$ and $-\pi < \theta \leqslant \pi$.
- $x = r\cos\theta$, $y = r\sin\theta$, $r = \sqrt{x^2 + y^2}$, $\theta = \arctan \dfrac{y}{x}$ ($\pm\pi$ if necessary).
- The area of a sector is $\int_\alpha^\beta \frac{1}{2} r^2 \, d\theta$.
- A conic is the locus of a point in a plane such that its distance from a fixed point S (the focus) is a constant multiple of its distance from a fixed line d (the directrix), both S and d being in the plane. The constant multiplier is the eccentricity e.
- The conic is an ellipse if $e < 1$, a parabola if $e = 1$, and a hyperbola if $e > 1$.
- The standard polar equation of a conic with focus as pole is $\dfrac{\ell}{r} = 1 + e\cos\theta$, where ℓ is the semi-latus rectum.

3 Calculus

The thought of the Differential Calculus warms my feet in bed.

Attributed to A.F. Pollard, 1869–1948

Differentiation of implicit functions

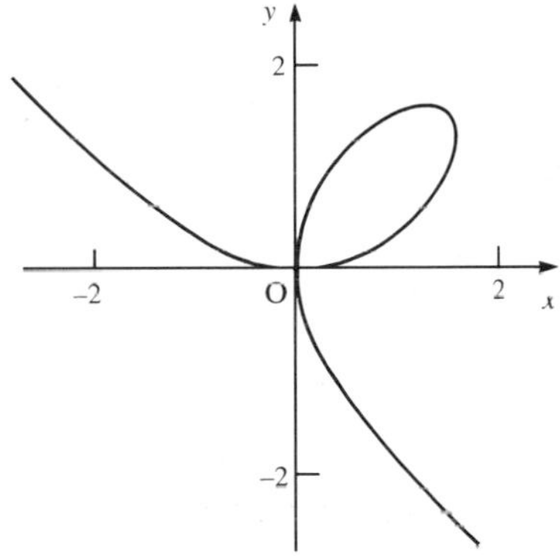

Figure 3.1

Figure 3.1 shows the graph of $x^3 + y^3 = 3xy$, known as the folium of Descartes. Can you find an expression for its gradient?

In your previous work you have dealt with functions defined explicitly by equations of the form $y = f(x)$. But you cannot make y the subject of the equation $x^3 + y^3 = 3xy$. When we specify a function by means of an equation connecting x and y which does not have y as the subject we are defining an implicit function. Another example is the equation $x^2 + y^2 = 1$; together with the restriction $y \geqslant 0$, this defines the function illustrated by the semicircle in figure 3.2; but $x^2 + y^2 = 1$, $y \leqslant 0$ clearly defines a different function. Although restrictions are often necessary to make the function unambiguous, we frequently assume that there are such restrictions but do not mention them.

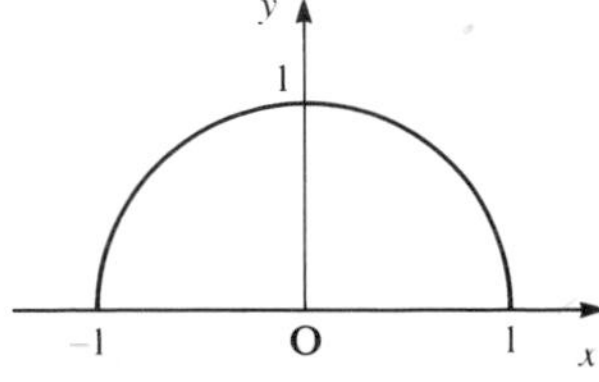

Figure 3.2

The chain rule, $\frac{dy}{dx}=\frac{dy}{du}\times\frac{du}{dx}$, enables us to differentiate (with respect to x) expressions such as y^3: $\frac{d}{dx}(y^3) = \frac{d}{dy}(y^3) \times \frac{dy}{dx} = 3y^2\frac{dy}{dx}$.

We will also make frequent use of the product rule:

$$\frac{d}{dx}(uv) = u\frac{dv}{dx} + v\frac{du}{dx}.$$

This means the derivative of the expression in the brackets with respect to x.

For example:

$$\frac{d}{dx}(xy) = x\frac{d}{dx}(y) + y\frac{d}{dx}(x) = x\frac{dy}{dx} + y.$$

EXAMPLE

Find an expression for $\frac{dy}{dx}$ for the folium of Descartes: $x^3 + y^3 = 3xy$. Hence find the co-ordinates of the stationary points.

Solution

$$x^3 + y^3 = 3xy \quad \Rightarrow \frac{d}{dx}(x^3) + \frac{d}{dx}(y^3) = \frac{d}{dx}(3xy)$$

Differentiating both sides with respect to x, term by term.

$$\Rightarrow 3x^2 + 3y^2\frac{dy}{dx} = 3\frac{d}{dx}(xy)$$

$$\Rightarrow 3x^2 + 3y^2\frac{dy}{dx} = 3\left(x\frac{dy}{dx} + y\right)$$

$$\Rightarrow y^2\frac{dy}{dx} - x\frac{dy}{dx} = y - x^2$$

Dividing by 3 and collecting terms involving $\frac{dy}{dx}$ on the same side.

$$\Rightarrow \frac{dy}{dx} = \frac{y - x^2}{y^2 - x}$$, an expression which involves both x and y.

Stationary points occur when $\frac{dy}{dx} = \frac{y-x^2}{y^2-x} = 0$, i.e. when $y = x^2$.

To find the co-ordinates we need to solve simultaneously $y = x^2$ and $x^3 + y^3 = 3xy$.

Substituting for y we obtain: $x^3 + x^6 = 3x^3 \Leftrightarrow x^6 - 2x^3 = 0 \Leftrightarrow x^3(x^3 - 2) = 0$.

The stationary points are $(0, 0)$ and $(\sqrt[3]{2}, \sqrt[3]{4})$.

(As interchanging x and y does not affect the equation $x^3 + y^3 = 3xy$ the graph is symmetrical about the line $y = x$, so that $(0, 0)$ and $(\sqrt[3]{4}, \sqrt[3]{2})$ are points where the tangents to the folium are parallel to the y axis.)

In the next example note carefully how we differentiate (with respect to x) various expressions involving y.

EXAMPLE

Differentiate each of the following with respect to x.

(i) x^3y^2 (ii) $(3x + 5y)^4$ (iii) $\ln y$ (iv) $(x^2 + y^3)\dfrac{dy}{dx}$

Solution

(i) $\dfrac{d}{dx}(x^3y^2) = x^3\dfrac{d}{dx}(y^2) + y^2\dfrac{d}{dx}(x^3)$ (product rule)

$= x^3\dfrac{d}{dy}(y^2)\dfrac{dy}{dx} + y^2(3x^2)$ (chain rule)

$= x^3(2y)\dfrac{dy}{dx} + 3x^2y^2 = 2x^3y\dfrac{dy}{dx} + 3x^2y^2$

(ii) $\dfrac{d}{dx}((3x + 5y)^4) = 4(3x + 5y)^3\dfrac{d}{dx}(3x + 5y)$ (chain rule)

$= 4(3x + 5y)^3\left(3 + 5\dfrac{dy}{dx}\right)$

(iii) $\dfrac{d}{dx}(\ln y) = \dfrac{d}{dy}(\ln y) \times \dfrac{dy}{dx} = \dfrac{1}{y}\dfrac{dy}{dx}$

(iv) $\dfrac{d}{dx}\left((x^2+y^3)\dfrac{dy}{dx}\right) = (x^2+y^3)\dfrac{d^2y}{dx^2} + \dfrac{d}{dx}(x^2+y^3) \times \dfrac{dy}{dx}$ (product rule)

$= (x^2+y^3)\dfrac{d^2y}{dx^2} + \left(2x + 3y^2\dfrac{dy}{dx}\right)\dfrac{dy}{dx}$

The next two examples again illustrate the finding of $\dfrac{dy}{dx}$ for implicit functions; the first also shows a typical method of finding $\dfrac{d^2y}{dx^2}$. Both examples show how the apparent laxity of not stating restrictions can sometimes be useful!

EXAMPLE

Find (i) $\dfrac{dy}{dx}$ and (ii) $\dfrac{d^2y}{dx^2}$ in terms of x and y given that $(x + 2)^2 + (y - 3)^2 = 25$.

Solution

(i) Differentiating both sides of $(x + 2)^2 + (y - 3)^2 = 25$ with respect to x gives

$\dfrac{d}{dx}[(x + 2)^2 + (y - 3)^2] = 0$ (Differentiating a constant gives 0.)

$\Rightarrow \dfrac{d}{dx}[(x + 2)^2] + \dfrac{d}{dy}[(y - 3)^2] = 0$

$\Rightarrow 2(x + 2) + 2(y - 3)\dfrac{dy}{dx} = 0$ (Applying chain rule to differentiate $(y - 3)^2$ with respect to y, and then converting differentiation to be with respect to x by multiplying by $\dfrac{dy}{dx}$.)

$\Rightarrow x + 2 + (y - 3)\dfrac{dy}{dx} = 0$

$\Rightarrow \dfrac{dy}{dx} = -\dfrac{x+2}{y-3}$.

(The equation represents the circle, centre $(-2, 3)$, radius 5. Inspection of the diagram shows that the gradient of CP = $\dfrac{y-3}{x+2}$; since the tangent at

P is perpendicular to CP, this confirms that $\frac{dy}{dx} = -\frac{x+2}{y-3}$. Notice that we are dealing with the two distinct functions

$(x + 2)^2 + (y - 3)^2 = 25, y \geqslant 3$ and $(x + 2)^2 + (y - 3)^2 = 25, y \leqslant 3$

and that the formula for the gradient is the same for both.)

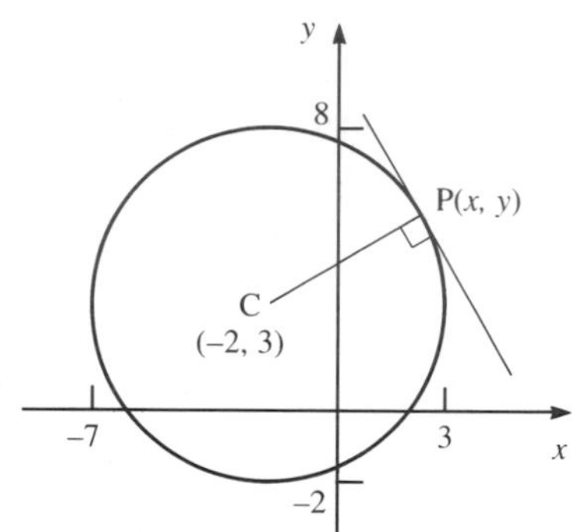

(ii) Having found an expression containing $\frac{dy}{dx}$ it is often easier to obtain $\frac{d^2y}{dx^2}$ by differentiating straight away without making $\frac{dy}{dx}$ the subject.

In (i) we showed that $x + 2 + (y - 3)\frac{dy}{dx} = 0$.

Differentiating with respect to x:

$$1 + \left((y-3)\frac{d^2y}{dx^2} + \left(\frac{dy}{dx}\right)^2\right) = 0$$

We differentiate $(y - 3)\frac{dy}{dx}$ as a product.

$$\Rightarrow (y-3)\frac{d^2y}{dx^2} = -\left(\frac{dy}{dx}\right)^2 - 1$$

$$= -\left(-\frac{x+2}{y-3}\right)^2 - 1$$

$$\Rightarrow \quad \frac{d^2y}{dx^2} = -\frac{(x+2)^2 + (y-3)^2}{(y-3)^3} = \frac{-25}{(y-3)^3} \text{ since } (x+2)^2 + (y-3)^2 = 25.$$

EXAMPLE

The equation of a curve is $4x^2 + 6xy = 9y^2 + 36$.

(i) Find the gradient of the curve at the points where $x = 3$.

(ii) Show that there are no stationary points.

(iii) Find the co-ordinates of the points where the tangent to the curve is parallel to the y axis.

Solution

$$4x^2 + 6xy = 9y^2 + 36$$

$$\Rightarrow 8x + \left(6y + 6x\frac{dy}{dx}\right) = 18y\frac{dy}{dx}$$

Differentiating term by term with respect to x and using the product to differentiate $6xy$.

$$\Rightarrow \frac{dy}{dx} = \frac{8x + 6y}{18y - 6x} = \frac{4x + 3y}{9y - 3x}.$$

(i) Substituting $x = 3$ in $4x^2 + 6xy = 9y^2 + 36$ gives $36 + 18y = 9y^2 + 36$

so that $9y^2 - 18y = 0 \Rightarrow y(y - 2) = 0 \Rightarrow y = 2$ or $y = 0$.

At (3, 2) the gradient is $\frac{4 \times 3 + 3 \times 2}{9 \times 2 - 3 \times 3} = 2$.

At (3, 0) the gradient is $\frac{4 \times 3 + 3 \times 0}{9 \times 0 - 3 \times 3} = -\frac{4}{3}$.

(ii) Stationary points occur when the numerator of $\frac{dy}{dx}$ is 0.

We need to solve simultaneously $4x + 3y = 0$ and $4x^2 + 6xy = 9y^2 + 36$:

substituting $-4x$ for $3y$ in the second equation gives

$$4x^2 + 2x(-4x) = (-4x)^2 + 36 \quad \Leftrightarrow 4x^2 - 8x^2 = 16x^2 + 36$$
$$\Leftrightarrow -20x^2 = 36$$

so that there is no real value of x for which $\frac{dy}{dx} = 0$.

There are no stationary points.

(iii) The tangent is parallel to the y axis when the denominator of $\frac{dy}{dx}$ is 0.

To find these points we solve simultaneously

$9y - 3x = 0$ and $4x^2 + 6xy = 9y^2 + 36$:

substituting $3y$ for x in the second equation gives

$$4(3y)^2 + 6(3y)y = 9y^2 + 36 \Leftrightarrow 36y^2 + 18y^2 = 9y^2 + 36$$
$$\Leftrightarrow 45y^2 = 36$$
$$\Leftrightarrow y = \pm \tfrac{2}{\sqrt{5}} \text{ and } x = \pm \tfrac{6}{\sqrt{5}}.$$

The tangents are parallel to the y axis at $(^6/\sqrt{5}, ^2/\sqrt{5})$ and at $(-^6/\sqrt{5}, -^2/\sqrt{5})$.

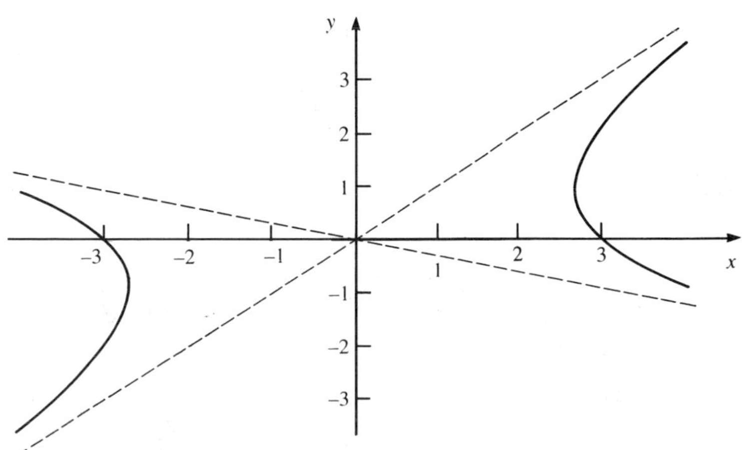

(The graph is the hyperbola shown in the diagram.)

Sometimes it is easier to find the derivative of products or quotients by first taking logarithms, and then differentiating as shown below, where we use y' to stand for $\frac{dy}{dx}$. (It is also common to use u' for $\frac{du}{dx}$, y'' for $\frac{d^2y}{dx^2}$, etc.)

EXAMPLE

Given that $y = \frac{(x^2-3)^2}{(x+1)}$ find y' in terms of x only.

Solution

$$y = \frac{(x^2-3)^2}{(x+1)}$$

$$\Rightarrow \quad \ln y = 2\ln(x^2-3) - \ln(x+1)$$

$$\Rightarrow \frac{1}{y} \times y' = \frac{2\times 2x}{x^2-3} - \frac{1}{x+1}$$

$$\Rightarrow \quad y' = y\left(\frac{4x}{x^2-3} - \frac{1}{x+1}\right)$$

$$= \frac{(x^2-3)^2}{(x+1)}\left(\frac{4x}{x^2-3} - \frac{1}{x+1}\right)$$

$$= \frac{(x^2-3)}{(x+1)^2}\left(4x(x+1) - (x^2-3)\right)$$

$$= \frac{(x^2-3)(3x^2+4x+3)}{(x+1)^2}$$

It is often unnecessary to substitute for y but in this example we want y' in terms of x only.

EXAMPLE

Use the fact that $\frac{d}{dx}(x^n) = nx^{n-1}$, where n is a positive integer, to prove that $y = x^n \Rightarrow y' = nx^{n-1}$ where n is any rational number.

Solution

First we prove that the formula holds when n is the negative integer $-p$:

$$y = x^{-p} \Rightarrow x^p y = 1.$$

Note that p is a positive integer.

Differentiating with respect to x: $px^{p-1}y + x^p y' = 0$

$$\Rightarrow y' = -px^{-1}y = -px^{-1}x^{-p} = -px^{-p-1} = nx^{n-1}.$$

Secondly we prove that the formula holds when n is the rational number $\frac{p}{q}$, where p and q are integers, $q \neq 0$: $\quad y = x^{p/q}$

$$\Rightarrow \quad y^q = x^p$$

$$\Rightarrow \quad qy^{q-1} \times y' = px^{p-1}$$

$$\Rightarrow \quad y' = \frac{px^{p-1}}{qy^{q-1}} = \frac{px^{p-1}y}{qy^q} = \frac{p}{q}\frac{x^{p-1}y}{x^p} = \frac{p}{q}\frac{y}{x}$$

$$= n\frac{x^n}{x} = nx^{n-1}.$$

Exercise 3A

1. Find $\frac{dy}{dx}$ and $\frac{d^2y}{dx^2}$ in terms of x and y:
[**Hint:** in (vi) and (vii) take logs first.]

(i) $x^3 + y^3 = 1$;

(ii) $xy = x + y$;

(iii) $(x - 3)(y + 7) = 5$;

(iv) $xy^2 = 2x + 3y$;

(v) $\sin x + \sin y = 1$;

(vi) $y = k^x$, where k is constant;

(vii) $y = x^x$.

2. Find the gradient of the curve $x^2 + 3xy + y^2 = x + y + 8$ at the point $(1, 2)$.

3. Find the co-ordinates of the stationary points on the curve $x^2 - xy + y^2 = 12$; find the value of y'' at each of these points, and so determine their nature.

4. Show that the graph of $xy + 48 = x^2 + y^2$ has stationary points at $(4, 8)$ and $(-4, -8)$, and find the co-ordinates of the points where the tangent is parallel to the y axis.

5. Find the co-ordinates of the stationary points on the curve $x^2 + y^2 + \frac{2x}{y} = 4$ and the co-ordinates of all points where the tangent to the curve is parallel to the y axis.

6. A curve C has equation $x^2 + y^2 + \frac{y}{x} = a$, where a is constant. Show that C has

(i) two points where the tangent is parallel to the y axis whatever the value of a;

(ii) two points where the tangent is parallel to the x axis provided a is positive.

If you have access to software which graphs implicit functions use it to sketch C for $a = 7$ and for $a = -0.3$.

7. In this question a, b, c are constants, and a and b are of opposite signs. Show that the graph of $ax^2 + xy + by^2 = c$ has tangents parallel to the x axis or the y axis, but does not have tangents parallel to both co-ordinate axes.

8. By differentiating $\ln(uvw)$ show that

$$\frac{d}{dx}(uvw) = u'vw + uv'w + uvw'.$$

9. Show that $\frac{d}{dx}\left(\frac{u}{vw}\right) = \frac{u'vw - uv'w - uvw'}{(vw)^2}$

and find a similar expression for $\frac{d}{dx}\left(\frac{st}{uvw}\right)$.

10. By taking logarithms of both sides first, find $\frac{dy}{dx}$ in terms of x only:

(i) $y = (2x - 3)^3(x + 4)^4$;

(ii) $y = \frac{3x - 7}{(5x + 2)^2}$;

(iii) $y = (2x^2 + 5)^3(x^2 + 3)^4$;

(iv) $y = \frac{(x^2 + 1)^2}{(x^2 - 3)^3}$;

(v) $y = \frac{(x^2 + 3)^3}{(x - 1)^2(2x + 1)^3}$;

(vi) $y = \sqrt{\frac{x^2 + 1}{x^3 + 2}}$.

11. The vapour pressure p and the absolute temperature T of mercury between 15°C and 270°C are connected by the equation $pT^c = Ae^{-b/T}$ where A, b, c are constants. Find $\frac{dp}{dT}$.

12. The sides a, b, c of triangle ABC are three consecutive terms of an arithmetic progression. Use the Sine Rule to show that $\sin A - 2\sin B + \sin C = 0$, and find $\frac{dA}{dB}$ in terms of cosines of A, B, C.

13. Find $\frac{dy}{dx}$ in terms of x, y, where $x^2 + y^2 = 2x + 2y$, and show that

$$\frac{d^2y}{dx^2} = \frac{2}{(1 - y)^3}.$$

14. The pressure p and volume V of a gas expanding adiabatically are related by the equation $pV^\gamma = c$, where γ, c are constants. Show that $V^2\frac{d^2p}{dV^2} = \gamma(\gamma + 1)p$ and find $p^2\frac{d^2V}{dp^2}$ in terms of γ, V only.

Exercise 3A continued

15. Van der Waal's equation

$$\left(p + \frac{a}{V^2}\right)(V - b) = RT$$

connects the pressure p, the volume V, and the temperature T of a gas, where a, b, R are constants. In this question take T as constant also. The critical point of a substance is when its gaseous and liquid forms have the same density: this occurs when $\frac{d^2p}{dV^2} = \frac{dp}{dV} = 0$. Find the values of p, V, T at the critical point in terms of a, b, R.

16. A particle of mass m is hung from a fixed point O by a light string of length ℓ. At time t the string makes a small angle θ with the vertical, and the mass has both potential energy P, and kinetic energy V, given by

$$P = mg\ell(1 - \cos\theta) \text{ and } V = \tfrac{1}{2}m\ell^2\left(\frac{d\theta}{dt}\right)^2$$

where m, g, and ℓ are constants. It is known that $P + V$ is constant, and observation shows that θ varies. By differentiating $P + V$ with respect to t show that $\frac{d^2\theta}{dt^2} \approx -\frac{g\theta}{\ell}$.

The inverse trigonometric functions

The arcsine function

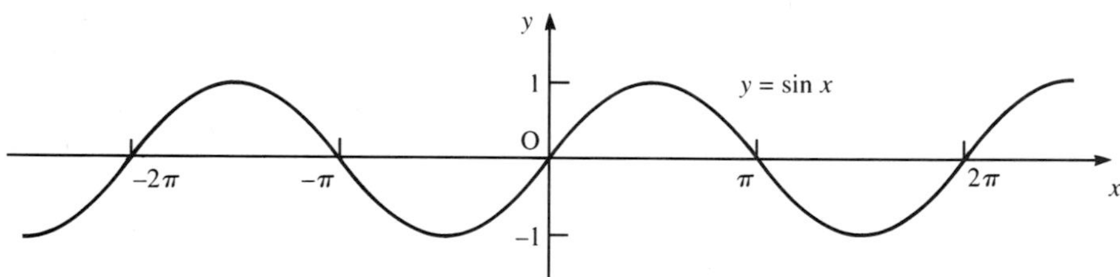

Figure 3.3

Figure 3.3 shows the graph of $y = \sin x$. The sine function is a many-to-one function: many values of x (for example: -2π, $-\pi$, 0, π, 2π, etc.) give the same value of y.

We can find the inverse of any function by interchanging x and y in the defining equation; this is equivalent to reflecting the graph in the line $y = x$. In the case of $y = \sin x$ we obtain the graph shown in figure 3.4; its equation is $x = \sin y$. For any value of x (between –1 and 1) there are infinitely many values of y, so figure 3.4 is not the graph of a function. However by restricting the range of y we can define a function, so that each value of x (between –1 and 1) is associated with a unique value of y. We can do this in infinitely many ways, but it is conventional (and sensible) to include $0 \leqslant y \leqslant \frac{\pi}{2}$ (i.e. angles in the first quadrant) as part of the required range, corresponding to $0 \leqslant x \leqslant 1$.

To keep our function continuous (and to have as large a domain as possible)we also include $-\frac{\pi}{2} \leqslant y < 0$, fourth quadrant angles, corresponding to $-1 \leqslant x < 0$. Figure 3.5 shows the complete graph of this function. Its equation is $y = \arcsin x$. (Older textbooks and many modern calculators use the notation $\sin^{-1}x$.) You will notice that the gradient of $y = \arcsin x$ is always positive, and that the gradient tends to infinity as $|x|$ tends to 1.

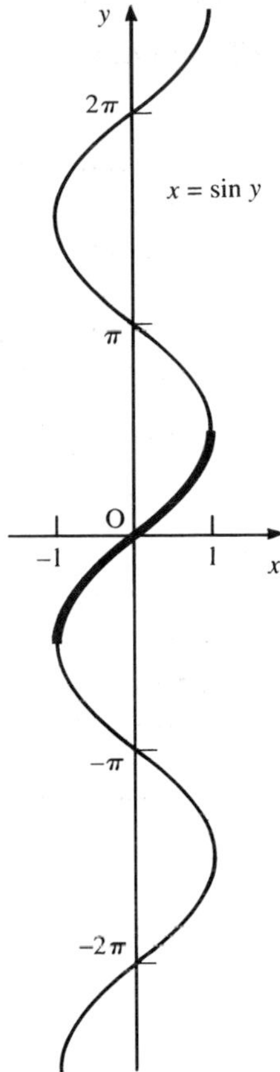

Now

$$
\begin{aligned}
& y = \arcsin x \\
\Rightarrow \quad & \sin y = x \\
\Rightarrow \quad & \frac{dy}{dx}\cos y = 1 \\
\Rightarrow \quad & \frac{dy}{dx} = \frac{1}{\cos y} \\
& \quad\;\; = \frac{1}{\pm\sqrt{1-\sin^2 y}} \\
& \quad\;\; = \frac{1}{\pm\sqrt{1-x^2}}.
\end{aligned}
$$

Figure 3.4

But $y = \arcsin x \Rightarrow -\frac{\pi}{2} \leqslant y \leqslant \frac{\pi}{2} \Rightarrow \cos y \geqslant 0$,

so that $\cos y = +\sqrt{1-x^2}$ and we conclude

that $\frac{d}{dx}(\arcsin x) = \frac{1}{\sqrt{1-x^2}}$.

Notice that the expression $\frac{1}{\sqrt{1-x^2}}$:

- is positive and only defined for $-1 < x < 1$,
- has a minimum at $x = 0$,
- tends to ∞ as x tends to ± 1,

all of which is consistent with the graph in figure 3.5.

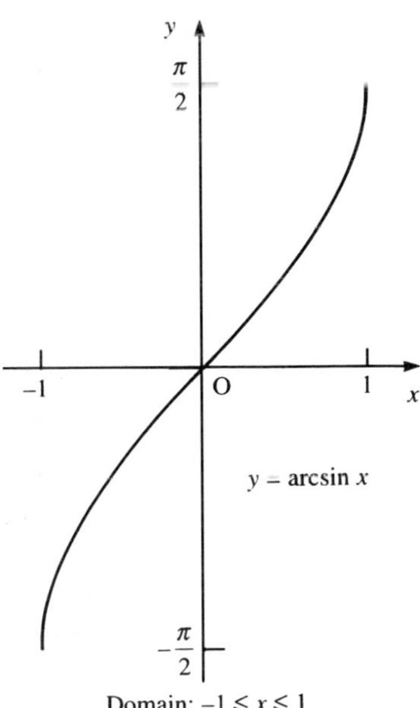

Domain: $-1 \leqslant x \leqslant 1$

Range: $-\frac{\pi}{2} \leqslant y \leqslant \frac{\pi}{2}$

Figure 3.5

The arccosine function

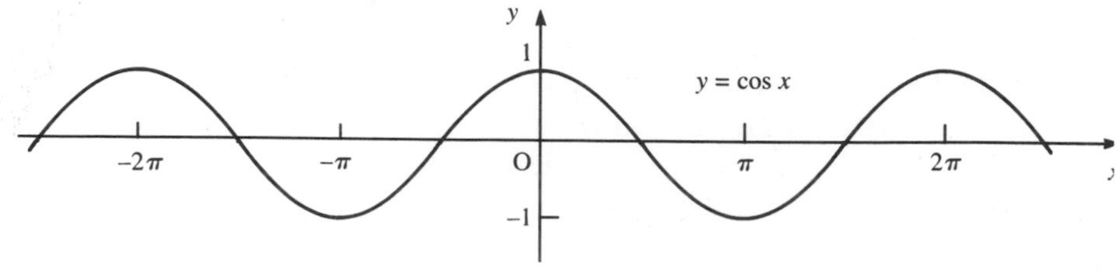

Figure 3.6

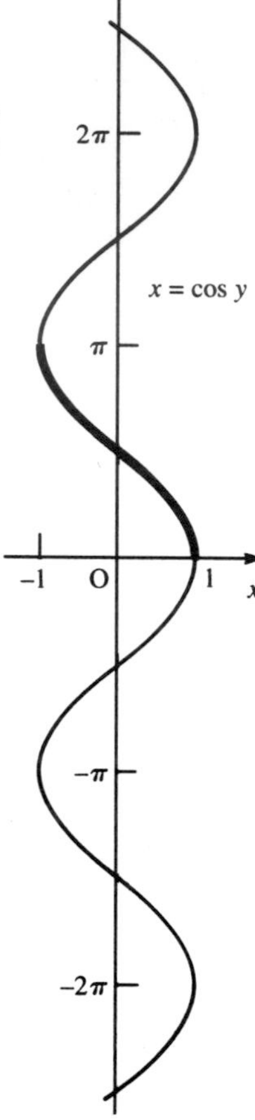

Figure 3.7

We deal with the inverse of the cosine function in much the same way. Figure 3.6 shows the graph of $y = \cos x$. Reflecting the graph of figure 3.6 in the line $y = x$ produces the graph with equation is $x = \cos y$ shown in figure 3.7. This is not the graph of a function. However by restricting the range of y we define a function, so that each value of x (between −1 and 1) is associated with a unique value of y. Again we choose to include $0 \leqslant y \leqslant \frac{\pi}{2}$ (first quadrant angles), corresponding to $0 \leqslant x \leqslant 1$. To preserve continuity and maximise the domain we also include $\frac{\pi}{2} < y \leqslant \pi$, second quadrant angles. Figure 3.8 shows the complete graph of this function. Its equation is $y = \arccos x$.

Activity

(i) From the various graphs (without using calculus) what can you say about the gradient of $y = \arccos x$?

(ii) Use calculus to show that

$$\frac{\mathrm{d}}{\mathrm{d}x}(\arccos x) = -\frac{1}{\sqrt{1-x^2}}.$$

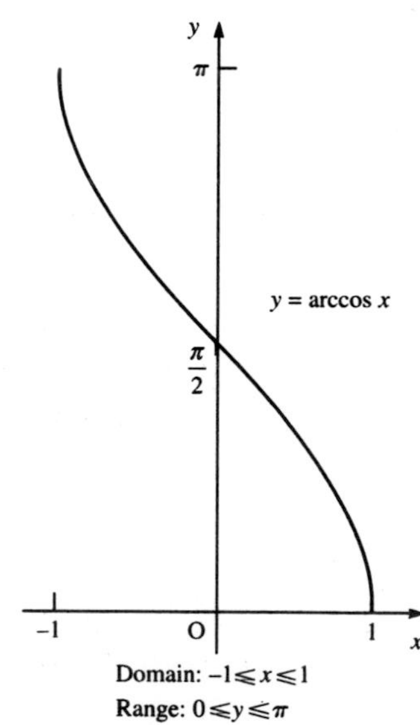

Figure 3.8

The arctangent function

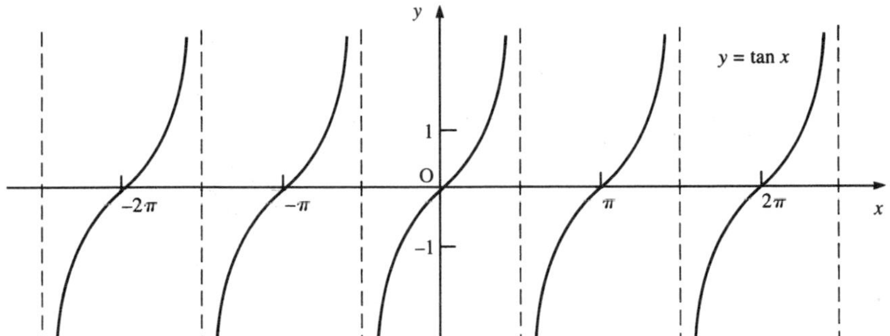

Figure 3.9

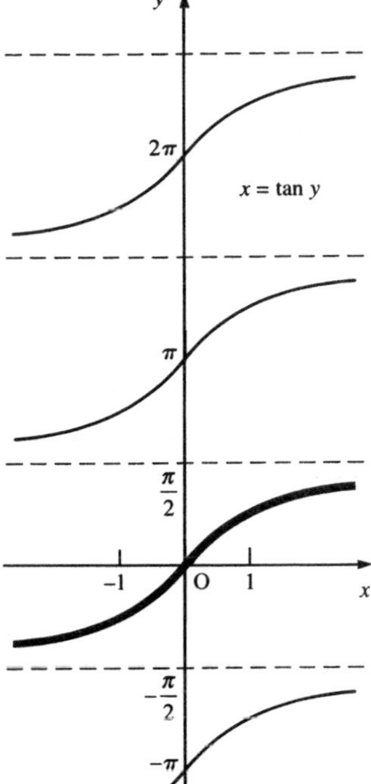

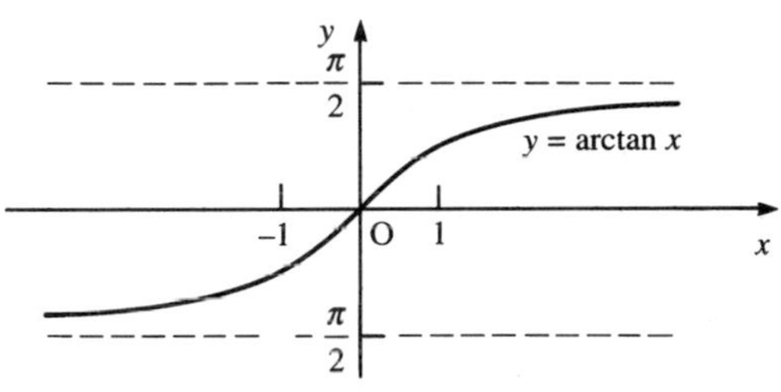

Domain: all real numbers

Range: $-\frac{\pi}{2} < y < \frac{\pi}{2}$

Figure 3.11

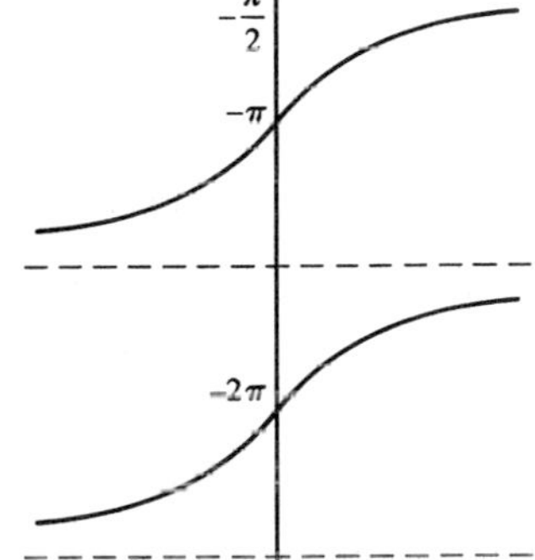

Figure 3.10

Activity

Describe the relationship of the various graphs shown in figures 3.9, 3.10 and 3.11.

Notice that the graph of $y = \arctan x$ has horizontal asymptotes. Describe qualitatively how its gradient varies and then use calculus methods to show that $\frac{d}{dx}(\arctan x) = \frac{1}{1+x^2}$.

The results $\frac{d}{dx}(\arcsin x) = \frac{1}{\sqrt{1-x^2}}$ and $\frac{d}{dx}(\arctan x) = \frac{1}{1+x^2}$ are particularly important, allowing you to integrate additional functions, as we shall show in the next section.

For Discussion

Show that (i) $\text{arcsec}\, x \equiv \arccos\left(\frac{1}{x}\right)$;

(ii) $\text{arccosec}\, x \equiv \arcsin\left(\frac{1}{x}\right)$;

(iii) $\text{arccot}\, x \equiv \arctan\left(\frac{1}{x}\right)$ with $\text{arccot}\, 0 = \frac{\pi}{2}$.

These formulae are useful as calculators, spreadsheets and other mathematical software frequently include only the three elementary trigonometric functions and their inverses.

General solutions

Solving various types of trigonometric equations was dealt with in *Pure Mathematics 3*, but solutions were generally restricted to a given range. If you are looking for the *general solution* of an equation you want a rule or formula which will give you all the solutions (and no other values).

For example: one solution of the equation $\sin x = \frac{1}{2}$ is $x = \arcsin \frac{1}{2} = \frac{\pi}{6}$; another solution is $x = \pi - \frac{\pi}{6} = \frac{5\pi}{6}$; all solutions are in the first or second quadrants as $\sin x$ is positive. Every solution may be regarded as a number of complete rotations plus $\frac{\pi}{6}$ or $\frac{5\pi}{6}$.

These two forms of solution may be written as

$x = 2m\pi + \frac{\pi}{6}$ or $(2m + 1)\pi - \frac{\pi}{6}$, where m is any integer.

Alternatively you may combine these two expressions and write the general solution of $\sin x = \frac{1}{2}$ as $x = n\pi + (-1)^n \frac{\pi}{6}$, where n is any integer.

The table shows the other important forms.

Equation	Form of the general solution (n is any integer)
$\sin x = y$	$x = n\pi + (-1)^n \arcsin y$
$\cos x = y$	$x = 2n\pi \pm \arccos y$
$\tan x = y$	$x = n\pi + \arctan y$

Exercise 3B

1. State the domain and range of the inverse sine, cosine and tangent functions.

2. Show that $\arcsin x + \arccos x \equiv \frac{\pi}{2}$.

3. Show that $\arcsin(-x) = -\arcsin x$, and that $\arctan(-x) = -\arctan x$. State and prove a formula connecting $\arccos(-x)$ and $\arccos x$.

4. Show that $\arcsin(\sin \pi) \neq \pi$. Under what circumstances is $\arcsin(\sin x) = x$?

5. Show that $\arccos \sqrt{1-x} \equiv \arcsin\sqrt{x}$.

6. Differentiate the following with respect to x.

(i) $\arcsin x$; (ii) $\arcsin 5x$;

(iii) $\arctan \frac{3x}{2}$; (iv) $\arctan(2 - 3x)$.

7. Differentiate the following with respect to x.

(i) $\arcsin 2x$ (ii) $\arctan 5x$

(iii) $\arcsin 3x^2$ (iv) $\arccos 2x$

(v) $\arctan(e^x)$ (vi) $\arctan(1 - x^2)$

(vii) $\arccos(5x^2 - 2)$ (viii) $\arcsin\sqrt{x}$

8. If $f(x) \equiv \sin x + \cos x$, $-\frac{\pi}{4} < x < \frac{\pi}{4}$, find $f^{-1}(x)$.

9. Write down the derivatives of $\arcsin x$ and $\arccos x$. Hence show that $\int \frac{1}{\sqrt{1-x^2}}\,dx$ may be expressed as $\arcsin x + c_1$ and as $\arccos x + c_2$, where c_1 and c_2 are arbitrary constants. Explain how the two results are compatible, and express c_2 in terms of c_1.

In questions 10 to 15 find the general solution of the equation; where possible give your answer as a rational multiple of π; otherwise leave your answer in a form involving an inverse trigonometric function.

10. $\sin 2x = \sin x$

11. $\cos x - \sin x = \sqrt{2}$

12. $3\cos x + 4\sin x = 2.5$

13. $\tan 2x = 4\tan x$

14. $\cos x = \cos \frac{1}{2}x + 1$

15. $2\sin x = \cos x + 1$

16. State the domain and range of

(i) $y = \operatorname{arcsec} x$;

(ii) $y = \operatorname{arccosec} x$;

(iii) $y = \operatorname{arccot} x$.

17. (i) (a) By sketching the graph of $y = \operatorname{arcsec} x$ show that $\frac{d}{dx}(\operatorname{arcsec} x) > 0$.

(b) Show that $\frac{d}{dx}(\operatorname{arcsec} x) = \frac{1}{|x|\sqrt{x^2-1}}$.

(ii) Find (a) $\frac{d}{dx}(\operatorname{arccosec} x)$;

(b) $\frac{d}{dx}(\operatorname{arccot} x)$.

18. Evaluate (i) $\operatorname{arcsec} x + \operatorname{arccosec} x$;

(ii) $\arctan x + \operatorname{arccot} x$.

Integration using inverse trigonometric functions

The inverse sine and tangent functions are particularly useful in integration.

Integration using the arcsine function

Since $\frac{d}{dx}(\arcsin x) = \frac{1}{\sqrt{1-x^2}}$ we know that $\int \frac{1}{\sqrt{1-x^2}}\,dx = \arcsin x + c$.

You will see the similarity between $\int \frac{1}{\sqrt{9-x^2}}\,dx$ and $\int \frac{1}{\sqrt{1-x^2}}\,dx$ and you may well (correctly) guess that $\int \frac{1}{\sqrt{9-x^2}}\,dx$ takes a similar form, but you will perhaps be unsure what effect the number 9 has on the expression. Try treating 9 as a factor:

$$\begin{aligned}\int \frac{1}{\sqrt{9-x^2}}\,\mathrm{d}x &= \int \frac{1}{\sqrt{9(1-\frac{x^2}{9})}}\,\mathrm{d}x \\ &= \int \frac{1}{3\sqrt{[1-(\frac{x}{3})^2]}}\,\mathrm{d}x \qquad \text{Let } 3u = x \text{ so that } 3\,\mathrm{d}u = \mathrm{d}x. \\ &= \int \frac{1}{3\sqrt{1-u^2}} \times 3\mathrm{d}u \\ &= \int \frac{1}{\sqrt{1-u^2}}\,\mathrm{d}u = \arcsin u + c = \arcsin \frac{x}{3} + c.\end{aligned}$$

We now construct the formula for $\int \frac{1}{\sqrt{a^2-x^2}}\,\mathrm{d}x$, where a is a positive constant.

As $x = 3u$ was a useful substitution when the denominator was $\sqrt{9-x^2}$, we use the substitution $x = au$ so that $\mathrm{d}x = a\,\mathrm{d}u$:

$$\begin{aligned}\int \frac{1}{\sqrt{a^2-x^2}}\,\mathrm{d}x &= \int \frac{1}{\sqrt{a^2-(au)^2}} \times a\,\mathrm{d}u \\ &= \int \frac{1}{a\sqrt{1-u^2}} \times a\,\mathrm{d}u \\ &= \int \frac{1}{\sqrt{1-u^2}}\,\mathrm{d}u = \arcsin u + c = \arcsin \frac{x}{a} + c.\end{aligned}$$

EXAMPLE

Find (i) $\int \frac{1}{\sqrt{16-x^2}}\,\mathrm{d}x$, (ii) $\int \frac{1}{\sqrt{16-3x^2}}\,\mathrm{d}x$.

Solution

(i) $\int \frac{1}{\sqrt{16-x^2}}\,\mathrm{d}x = \arcsin \frac{x}{4} + c.$

This is of the form $\int \frac{1}{\sqrt{a^2-x^2}}\,\mathrm{d}x$ with $a = 4$.

(ii) $\int \frac{1}{\sqrt{16-3x^2}}\,\mathrm{d}x = \frac{1}{\sqrt{3}}\int \frac{1}{\sqrt{\frac{16}{3}-x^2}}\,\mathrm{d}x = \frac{1}{\sqrt{3}}\arcsin \frac{x\sqrt{3}}{4} + c.$

Take out the factor $\sqrt{3}$, then as in (i) with $a = \frac{4}{\sqrt{3}}$.

Integration using the arctangent function

In the same way knowing that $\frac{\mathrm{d}}{\mathrm{d}x}(\arctan x) = \frac{1}{1+x^2}$ so that

$\int \frac{1}{1+x^2}\,\mathrm{d}x = \arctan x + c$ may well lead you to guess that $\int \frac{1}{a^2+x^2}\,\mathrm{d}x$ takes a similar form. But

$$\begin{aligned}\int \frac{1}{a^2+x^2}\,\mathrm{d}x &= \int \frac{1}{a^2+a^2u^2} \times a\,\mathrm{d}u \qquad \text{putting } au = x \text{ so that } a\,\mathrm{d}u = \mathrm{d}x \\ &= \int \frac{1}{a(1+u^2)}\,\mathrm{d}u \\ &= \frac{1}{a}\arctan u + c = \frac{1}{a}\arctan \frac{x}{a} + c.\end{aligned}$$

Notice the factor $\frac{1}{a}$.

EXAMPLE

Find (i) $\int \frac{1}{5+x^2}\,dx$; (ii) $\int \frac{1}{5+4x^2}\,dx$.

Solution

(i) $\int \frac{1}{5+x^2}\,dx = \frac{1}{\sqrt{5}}\arctan\frac{x}{\sqrt{5}} + c.$

This is of the form $\int \frac{1}{a^2+x^2}\,dx$ with $a = \sqrt{5}$.

(ii) $\int \frac{1}{5+4x^2}\,dx = \frac{1}{4}\int \frac{1}{\frac{5}{4}+x^2}\,dx = \frac{1}{4} \times \frac{1}{\sqrt{\frac{5}{4}}}\arctan\frac{1}{\sqrt{\frac{5}{4}}} + c = \frac{1}{2\sqrt{5}}\arctan\frac{2x}{\sqrt{5}} + c.$

Take out the factor 4, then as in (i) with $a = \sqrt{\frac{5}{4}}$.

NOTE

Dimensions will help you understand (and remember) why the factor $\frac{1}{a}$ *is needed in*

$$\int \frac{1}{a^2+x^2}\,dx = \frac{1}{a}\arctan\frac{x}{a} + c \qquad ①$$

but not in

$$\int \frac{1}{\sqrt{a^2-x^2}}\,dx = \arcsin\frac{x}{a} + c. \qquad ②$$

Integration is a form of summation. In both integrals dx *is a length. In ② the expression* $\int \frac{1}{\sqrt{a^2-x^2}}$ *is a number divided by the square root of an area; multiplying by* dx *gives a dimensionless number; the sum of a series of numbers is dimensionless;* $\arcsin\frac{x}{a}$ *is an angle, also dimensionless. So ② is dimensionally correct. In ① the expression* $\frac{1}{a^2+x^2}$ *is a number divided by an area; multiplying by* dx *gives the dimension* L^{-1} *(i.e. the reciprocal of a length); summing these does not change the dimension;* $\arctan\frac{x}{a}$ *is dimensionless and multiplying it by something like* $\frac{1}{a}$ *(with the dimension* L^{-1}*) makes the two sides of ① agree dimensionally. (In ① the constant c has the dimension* L^{-1}*; in ② the constant c is dimensionless.)*

The next example involves definite integration.

EXAMPLE

Evaluate $\int_0^2 \frac{1}{4+x^2}\,dx$.

Solution

$$\int_0^2 \frac{1}{4+x^2}\,dx = \left[\frac{1}{2}\arctan\frac{x}{2}\right]_0^2 = \frac{1}{2}(\arctan 1 - \arctan 0) = \frac{\pi}{8}$$

Alternative approach

Alternatively we may make the substitution $x = 2\tan u$, remembering to change the limits of integration at the same time. But the equation $x = 2\tan u$ does not define u uniquely: given $x = 0$, for example, u may be 0, or π, or any multiple of π. However, though it looks more cumbersome, the equation

$u = \arctan \frac{x}{2}$ does define u uniquely, and is our preferred way of stating the substitution. Then

$$\int_0^2 \frac{1}{4+x^2}\,dx = \int_0^{\frac{\pi}{4}} \frac{2\sec^2 u}{4\sec^2 u}\,du = \int_0^{\frac{\pi}{4}} \tfrac{1}{2}\,du = \tfrac{1}{2}[u]_0^{\frac{\pi}{4}} = \frac{\pi}{8}.$$

where $u = \arctan \frac{x}{2}$ $\Rightarrow x = 2\tan u$ $\Rightarrow dx = 2\sec^2 u\,du.$

When $x = 2$, $u = \frac{\pi}{4}$; when $x = 0$, $u = 0$.

Exercise 3C

1 – 6. Find the indefinite integrals.

1. $\int \frac{1}{25+x^2}\,dx$

2. $\int \frac{1}{\sqrt{36-x^2}}\,dx$

3. $\int \frac{5}{x^2+36}\,dx$

4. $\int \frac{4}{25+4x^2}\,dx$

5. $\int \frac{1}{\sqrt{9-4x^2}}\,dx$

6. $\int \frac{7}{\sqrt{5-3x^2}}\,dx$

7 – 12. Evaluate the definite integrals, leaving your answers in terms of π.

7. $\int_0^3 \frac{1}{9+x^2}\,dx$

8. $\int_0^{\sqrt{2}} \frac{1}{\sqrt{4-x^2}}\,dx$

9. $\int_{-\frac{1}{\sqrt{3}}}^{\frac{1}{3}} \frac{1}{1+9x^2}\,dx$

10. $\int_0^{\frac{1}{4}} \frac{1}{\sqrt{1-4x^2}}\,dx$

11. $\int_{-\frac{1}{2}}^{\frac{1}{2}} \frac{1}{\sqrt{3-6x^2}}\,dx$

12. $\int_{\sqrt{\frac{5}{6}}}^{\sqrt{\frac{5}{2}}} \frac{1}{5+2x^2}\,dx$

Harder integrations

You have been integrating functions of the form: $\frac{1}{a^2+x^2}$ and $\frac{1}{\sqrt{a^2-x^2}}$.

The example below shows how the formula $\int \frac{1}{a^2+x^2}\,dx = \frac{1}{a}\arctan\frac{x}{a} + c$ helps us integrate rational functions with constant numerator, and a denominator which is quadratic with no real roots.

EXAMPLE

Find $\int \frac{4}{x^2 - 2x + 3}\,dx$.

Solution

$$\int \frac{4}{x^2 - 2x + 3}\,dx = 4\int \frac{1}{(x-1)^2 + 2}\,dx$$

Taking out the factor 4, and 'completing the square' in the denominator.

$$= 4\int \frac{1}{u^2 + 2}\,du \quad \text{where } u = x - 1 \text{ and } du = dx$$

$$= 4 \times \frac{1}{\sqrt{2}} \arctan \frac{u}{\sqrt{2}} + c$$

You may well be able to omit these two lines.

$$= 2\sqrt{2} \arctan \frac{x-1}{\sqrt{2}} + c.$$

For Discussion

When trying to integrate $\frac{1}{Ax^2 + Bx + C}$ how can we tell if $Ax^2 + Bx^2 + C$ has no real roots? And how should we proceed if A is negative?

The next example shows how the formula $\int \frac{1}{\sqrt{a^2 - x^2}}\,dx = \arcsin \frac{x}{a} + c$ helps us integrate functions that can be arranged as a fraction, with constant numerator, and a denominator which is the square root of a quadratic; this quadratic must have distinct real roots and the coefficient of x^2 must be negative.

EXAMPLE

Find $\int \frac{5}{\sqrt{2 + 4x - 4x^2}}\,dx$.

Solution

$$\int \frac{5}{\sqrt{2 + 4x - 4x^2}}\,dx = \frac{5}{2}\int \frac{1}{\sqrt{\frac{1}{2} - (x^2 - x)}}\,dx$$

Taking out the factor 5 from the numerator, and the factor 4 from 'inside' the square root in the denominator.

$$= \frac{5}{2}\int \frac{1}{\sqrt{\frac{3}{4} - (x - \frac{1}{2})^2}}\,dx$$

Completing the square first and then adjusting the constant to get $\frac{3}{4}$.

$$= \frac{5}{2}\int \frac{1}{\sqrt{\frac{3}{4} - u^2}}\,du \quad \text{where } u = x - \tfrac{1}{2} \text{ and } du = dx$$

$$= \frac{5}{2} \arcsin \frac{2u}{\sqrt{3}} + c$$

Again you may well be able to omit these two lines.

$$= \frac{5}{2} \arcsin \frac{2x - 1}{\sqrt{3}} + c.$$

For Discussion

When using the formula $\int \frac{1}{\sqrt{a^2-x^2}}\mathrm{d}x = \arcsin\frac{x}{a} + c$ to integrate $\frac{1}{\sqrt{Ax^2+Bx+C}}$ why is it necessary to have A negative and $B^2 > 4AC$?

The final example illustrates other ways these techniques may be used.

EXAMPLE

Find (i) $\int \frac{x+5}{x^2+4}\mathrm{d}x$; (ii) $\int \frac{x}{\sqrt{1-x^2}}\mathrm{d}x$; (iii) $\int \frac{9x-8}{(x^2+9)(x+2)}\mathrm{d}x$.

Solution

(i)
$$\begin{aligned}\int \frac{x+5}{x^2+4}\mathrm{d}x &= \int\left(\frac{x}{x^2+4}+\frac{5}{x^2+4}\right)\mathrm{d}x \\ &= \frac{1}{2}\int\frac{2x}{x^2+4}\mathrm{d}x + 5\int\frac{1}{x^2+4}\mathrm{d}x \\ &= \frac{1}{2}\ln\left(x^2+4\right)+5\times\frac{1}{2}\arctan\frac{x}{2}+c \\ &= \frac{1}{2}\ln\left(x^2+4\right)+\frac{5}{2}\arctan\frac{x}{2}+c\,.\end{aligned}$$

The fraction being integrated is split into two parts: one numerator = constant × derivative of denominator; the other numerator is constant.

(ii) $\int \frac{x}{\sqrt{1-x^2}}\mathrm{d}x$ is best found by inspection:

$$\frac{\mathrm{d}}{\mathrm{d}x}\left(1-x^2\right)^{\frac{1}{2}} = \tfrac{1}{2}\left(1-x^2\right)^{-\frac{1}{2}}\times(-2x) = -\frac{x}{\sqrt{1-x^2}}$$

Alternatively use any of the substitutions $u = \arcsin x$; $u = 1 - x^2$; $u^2 = 1 - x^2$.

so that $\int \frac{x}{\sqrt{1-x^2}}\mathrm{d}x = -\sqrt{1-x^2+c}\,.$

(iii)
$$\begin{aligned}\int \frac{9x-8}{(x^2+9)(x+2)}\mathrm{d}x &= \int\left(\frac{2x+5}{x^2+9}-\frac{2}{x+2}\right)\mathrm{d}x \\ &= \int\frac{2x+5}{x^2+9}\mathrm{d}x - \int\frac{2}{x+2}\mathrm{d}x \\ &= \int\frac{2x}{x^2+9}\mathrm{d}x + \int\frac{5}{x^2+9}\mathrm{d}x - \int\frac{2}{x+2}\mathrm{d}x \\ &= \ln\left(x^2+9\right)+5\times\frac{1}{3}\arctan\frac{x}{3} - 2\ln|x+2| + c \\ &= \ln\frac{x^2+9}{(x+2)^2}+\frac{5}{3}\arctan\frac{x}{3}+c.\end{aligned}$$

The rational function being integrated is expressed in partial fractions: see *Pure Mathematics 3*.

As $x^2 + 9$ and $(x + 2)^2$ are clearly positive we do not need to use modulus signs here.

Exercise 3D

1. Find the following integrals.

(i) $\int \frac{1}{4+(x+2)^2}\,dx$

(ii) $\int \frac{7}{\sqrt{5+4x-x^2}}\,dx$

(iii) $\int \frac{3}{3+2x^2}\,dx$

(iv) $\int \frac{3}{9x^2+6x+5}\,dx$

(v) $\int \frac{1}{\sqrt{3+2x-x^2}}\,dx$

(vi) $\int \frac{7}{\sqrt{3-4x-4x^2}}\,dx$

2. (i) By writing $\arcsin x$ as $1\times\arcsin x$ use integration by parts to find $\int \arcsin x\,dx$.

(ii) Use a similar method to find the following integrals.

(a) $\int \arccos x\,dx$

(b) $\int \arctan x\,dx$

(c) $\int \operatorname{arccot} x\,dx$

3. (i) Use the substitution $x = a\sin u$ to find $\int_0^b \sqrt{(a^2-x^2)}\,dx$, where $a>b>0$.

(ii) Draw a sketch to show the significance of the area you calculated in 3 (i), and explain both terms of your answer to (i) geometrically.

4. Find the following integrals.

(i) $\int \frac{1}{x^2-6x+13}\,dx$

(ii) $\int \frac{1}{\sqrt{7-12x-4x^2}}\,dx$

(iii) $\int \frac{1}{4x^2+20x+29}\,dx$

(iv) $\int \frac{1}{x^2-6x+9}\,dx$

(v) $\int \frac{1}{\sqrt{5-12x-9x^2}}\,dx$

5. Find the following integrals.

(i) $\int \frac{x+1}{x^2+1}\,dx$

(ii) $\int \frac{4}{(x^2+1)(1+x)}\,dx$

(iii) $\int \frac{1-x}{\sqrt{1-x^2}}\,dx$

(iv) $\int \frac{x+3}{(x+1)(x^2+1)}\,dx$

6. Evaluate the following:

(i) $\int_1^3 \frac{1}{\sqrt{4x-x^2}}\,dx$

(ii) $\int_2^5 \frac{2x^2+3}{(x-1)(x^2+4)}\,dx$

7. Find $\frac{d}{dx}(\operatorname{arcsec} x)$ and $\int \frac{dx}{x\sqrt{x^2-a^2}}$.

Maclaurin series

In this section you will learn a technique which allows you to express a function f(x) as the sum of an infinite series. But we start by looking at the related question of forming a sequence of polynomials (of increasing degree) that successively approximate to f(x).

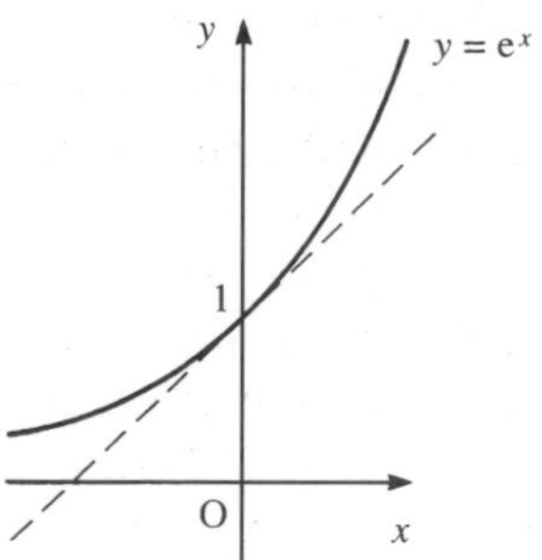

Figure 3.12

If you want to use a straight line to approximate the curve with equation $y = e^x$, there are many straight lines you could choose. Even restricting your choice to those which are tangents, there are infinitely many lines you could choose. For simplicity we use the straight line which is tangent to the curve at the point where $x = 0$, as illustrated in figure 3.12. Suppose the tangent has equation $y = a_0 + a_1x$; then:

① line and curve cut y axis at same point $\Rightarrow a_0 = e^0 = 1$;

② line and curve have the same gradient when $x = 0 \Rightarrow a_1 = 1$ since $\frac{d}{dx}\left(e^x\right) = e^x$, which is 1 when $x = 0$.

So the linear approximation for e^x is $1 + x$.

But straight lines are straight, and are not really suitable for approximating to curves over any distance. Using the quadratic equation, $y = a_0 + a_1x + a_2x^2$ to approximate to $y = e^x$, as shown in figure 3.13, requires as before, that:

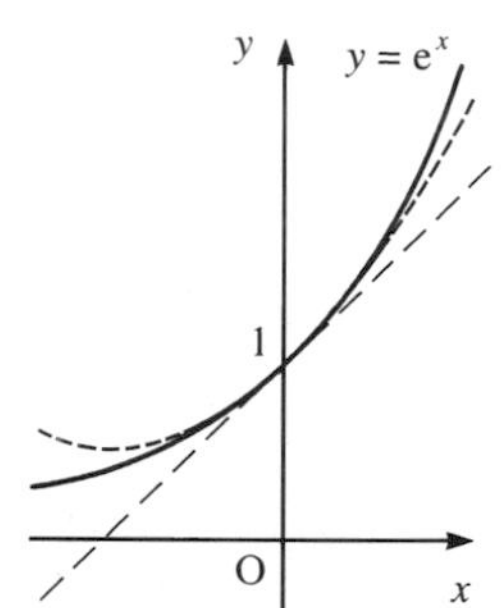

Figure 3.13

① both curves cut the y axis at same point $\Rightarrow a_0 = e^0 = 1$;

② both curves have the same gradient when $x = 0 \Rightarrow a_1 = e^0 = 1$;

but now: ③ both curves must have the same second derivative when $x = 0$.

Since $\frac{d^2}{dx^2}\left(e^x\right) = e^x$, which is 1 when $x = 0$, and $\frac{d^2}{dx^2}\left(a_0 + a_1x + a_2x^2\right) = 2a_2$,

③ $\Rightarrow 2a_2 = 1 \Rightarrow a_2 = \frac{1}{2}$. So the quadratic approximation for e^x is $1 + x + \frac{1}{2}x^2$.

Extending this to finding the cubic $a_0 + a_1x + a_2x^2 + a_3x^3$ that approximates to e^x brings in the additional requirement that the cubic and e^x have the same third derivative at $x = 0$. Now $\frac{d^3}{dx^3}(a_0 + a_1x + a_2x^2 + a_3x^3) = 3 \times 2a_3 = 3!\, a_3$ and

$\frac{d^3}{dx^3}\left(e^x\right) = e^x = 1$ when $x = 0$ so we require $a_3 = \frac{1}{3!}$. The cubic approximation for e^x is $1 + x + \frac{1}{2!}x^2 + \frac{1}{3!}x^3$.

Figure 3.14 shows the graph of $y = e^x$ together with the graphs of the linear, quadratic and cubic approximations we have just constructed. The graph shows that, for positive x, the accuracy of the approximation improves as more terms are used. Using more terms also improves the accuracy when x is negative, though the diagram alone does not justify that claim.

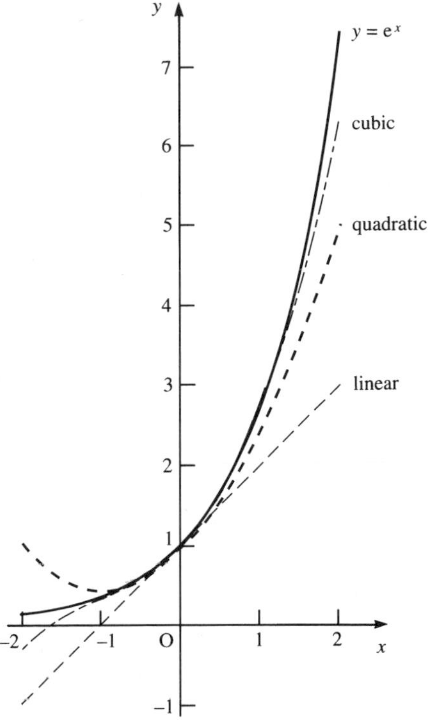

Figure 3.14

Activity

The cubic approximation for e^x is $1 + x + \frac{x^2}{2} + \frac{x^3}{6}$. Write down a cubic approximation for e^{-x}. Multiply the two approximations together and comment on your answer.

Activity

Show that the next (i.e. the fourth degree) approximation for e^x is $1 + x + \frac{x^2}{2!} + \frac{x^3}{3!} + \frac{x^4}{4!}$.

This is typical of the general case. Suppose $f(x)$ is a function, that its first n derivatives exist at $x = 0$, and that you want to find a polynomial $p(x)$ of degree n which has the same values as $f(x)$ and its first n derivatives at $x = 0$. Then

$$p(0) = f(0),\ p'(0) = f'(0),\ p''(0) = f''(0),\ \ldots,\ p^{(n)}(0) = f^{(n)}(0).$$

Note the symbol for the nth derivative of $p(x)$, evaluated at $x = 0$.

Solving these $n + 1$ equations gives the $n + 1$ coefficients needed for a polonomial of degree n.

If $p(x) \equiv a_0 + a_1x + a_2x^2 + a_3x^3 + \ldots + a_rx^r + \ldots + a_nx^n$ then $p(0) = a_0 = f(0)$ and

$p'(x) \equiv a_1 + 2a_2x + 3a_3x^2 + \ldots + ra_rx^{r-1} + \ldots + na_nx^{n-1} \quad \Rightarrow p'(0) = a_1 = f'(0);$

$p''(x) \equiv 2a_2 + 6a_3x + \ldots + r(r-1)a_rx^{r-2} + \ldots + n(n-1)a_nx^{n-2} \quad \Rightarrow p''(0) = 2a_2 = f''(0);$

$p'''(x) \equiv 6a_3 + \ldots + r(r-1)(r-2)a_rx^{r-3} + \ldots + n(n-1)(n-2)a_nx^{n-3} \Rightarrow p'''(0) = 6a_3 = f'''(0);$

and so on. Generalising:

$p^{(r)}(x) \equiv r!a_r + \ldots + n(n-1)(n-2)\ldots(n-r+1)a_nx^{n-r} \quad \Rightarrow p^{(r)}(0) = r!a_r = f^{(r)}(0);$

and $p^{(n)}(x) \equiv n!a_n \quad \Rightarrow p^{(n)}(0) = n!a_n = f^{(n)}(0).$

From these we find the coefficients a_0, a_1, a_2, etc., and conclude that

$$f(x) \approx f(0) + x\,f'(0) + \frac{x^2}{2!}f''(0) + \frac{x^3}{3!}f'''(0) + \ldots + \frac{x^r}{r!}f^{(r)}(0) + \ldots + \frac{x^n}{n!}f^{(n)}(0).$$

This is known as the *Maclaurin expansion* for f(x) as far as the term in x^n, or the nth *Maclaurin approximation* for f(x).

EXAMPLE

Find the Maclaurin expansion for $(1-x)^{-1}$ as far as x^n.

Solution

Let $f(x) \equiv (1-x)^{-1}$.

$f(x) \equiv (1-x)^{-1}$	$f(0) = 1$
$f'(x) \equiv (1-x)^{-2}$	$f'(0) = 1$
$f''(x) \equiv 2(1-x)^{-3}$	$f''(0) = 2$
$f'''(x) \equiv 6(1-x)^{-4}$	$f'''(0) = 6$
$f^{(iv)}(x) \equiv 24(1-x)^{-5}$	$f^{(iv)}(0) = 24$
...	...
$f^{(n)}(x) \equiv n!(1-x)^{-(n+1)}$	$f^{(n)}(0) = n!$

Tabulate f(x) and its derivatives and evaluate them at $x = 0$.

Then $(1-x)^{-1} \approx 1 + x + x^2 + x^3 + x^4 + \ldots + x^n$.

Validity

At this stage we cannot say much about the accuracy of these Maclaurin approximations. But the nth Maclaurin expansion for $(1-x)^{-1}$, just obtained, is the geometric progression $1 + x + x^2 + x^3 + x^4 + \ldots + x^n$; if we let n tend to infinity we obtain the infinite geometric series $1 + x + x^2 + x^3 + x^4 + \ldots$ which, if $|x| < 1$, converges to $\frac{1}{1-x} \equiv (1-x)^{-1}$, known as its *sum to infinity*. This means that, provided $|x| < 1$, by taking sufficiently many terms we can make the Maclaurin expansion of $(1-x)^{-1}$ as close to $(1-x)^{-1}$ as we like. But the geometric series $1 + x + x^2 + x^3 + x^4 + \ldots$ does not converge if $|x| \geqslant 1$.

Generalising the ideas above: if the function f(x) and all its derivatives exist at $x = 0$, then the infinite series $f(0) + x\,f'(0) + \frac{x^2}{2!}f''(0) + \frac{x^3}{3!}f'''(0) + \ldots + \frac{x^r}{r!}f^{(r)}(0) + \ldots$ is known as the *Maclaurin series* for f(x). If the sum of this series up to and including the term in x^n (i.e. the sum of the first $n + 1$ terms) tends to a limit as n tends to infinity, and this limit is f(x), we say that the expansion *converges* to f(x). For some functions, for example $(1-x)^{-1}$, the series only converges for a limited range of values of x; we describe these values as those for which the series is *valid*. A more detailed examination of the validity of the Maclaurin series will be given in *Pure Mathematics 6*, but for now we merely state, without proof, the values of x for which the common Maclaurin series are valid. Though rare, there are examples in which the Maclaurin series for $f(x)$ converges, but not to $f(x)$.

We started this section by developing Maclaurin expansions for e^x. Since e^x and all its derivatives are identical, and $e^x = 1$ when $x = 0$, the Maclaurin series for e^x is $1 + x + \frac{x^2}{2!} + \frac{x^3}{3!} + \ldots + \frac{x^r}{r!} + \ldots$. This series is valid for all x.

EXAMPLE

Find the Maclaurin series for $\sin x$.

Solution

Let $f(x) \equiv \sin x$.

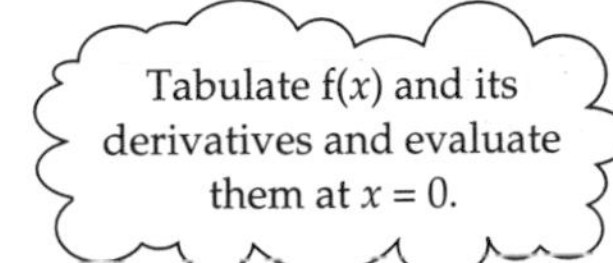

$f(x) \equiv \sin x$	$f(0) = 0$
$f'(x) \equiv \cos x$	$f'(0) = 1$
$f''(x) \equiv -\sin x$	$f''(0) = 0$
$f'''(x) \equiv -\cos x$	$f'''(0) = -1$
$f^{(iv)}(x) \equiv \sin x$	$f^{(iv)}(0) = 0$

$f^{(2r+1)}(x) \equiv (-1)^r \cos x$	$f^{(2r+1)}(0) = (-1)^r$
$f^{(2r+2)}(x) \equiv (-1)^{r+1} \sin x$	$f^{(2r+2)}(0) = 0$
$f^{(2r+3)}(x) \equiv (-1)^{r+1} \cos x$	$f^{(2r+3)}(0) = (-1)^{r+1}$
$f^{(2r+4)}(x) \equiv (-1)^{r+2} \sin x$	$f^{(2r+4)}(0) = 0$

Note the connection between these terms and the fact that $\sin x$ is an odd function.

Then $\sin x = x - \frac{x^3}{3!} + \frac{x^5}{5!} - \frac{x^7}{7!} + \ldots + \frac{(-1)^r x^{2r+1}}{(2r+1)!} + \ldots$.
(This series is valid for all values of x.)

Activity

Show that the Maclaurin series for $\cos x$ is $1 - \frac{x^2}{2!} + \frac{x^4}{4!} - \frac{x^6}{6!} + \ldots + \frac{(-1)^r x^{2r}}{(2r)!} + \ldots$

(This series is also valid for all x. Notice that the first two terms here form the familiar approximation for $\cos x$ when x is small, and that, as you might expect, the series for $\cos x$ is the same as the series obtained by differentiating the $\sin x$ series term by term.)

Activity

Show that the Maclaurin series for $(1 + x)^n$ is

$$1 + nx + \frac{n(n-1)}{2!}x^2 + \ldots + \frac{n(n-1)\ldots(n-r+1)}{r!}x^r + \ldots$$

i.e. the familiar binomial series for $(1 + x)^n$.

If n is a positive integer: the series terminates after $n + 1$ terms, and is valid for all x.

If n is not a positive integer: the series is valid for $|x| < 1$, but not valid for $|x| > 1$; the series is also valid for $x = 1$ if $n > -1$, and for $x = -1$ if $n > 0$.

Figure 3.15 shows the graph of the function $(1+x)^{-\frac{1}{2}}$ and several successive Maclaurin approximations. It illustrates the fact that the approximations converge on $(1+x)^{-\frac{1}{2}}$ if $|x| < 1$, but not if $x > 1$. At first sight the graph may appear to show that successive approximations also converge when $x < -1$; but they cannot be converging on $(1+x)^{-\frac{1}{2}}$, which is undefined for $x \leqslant -1$.

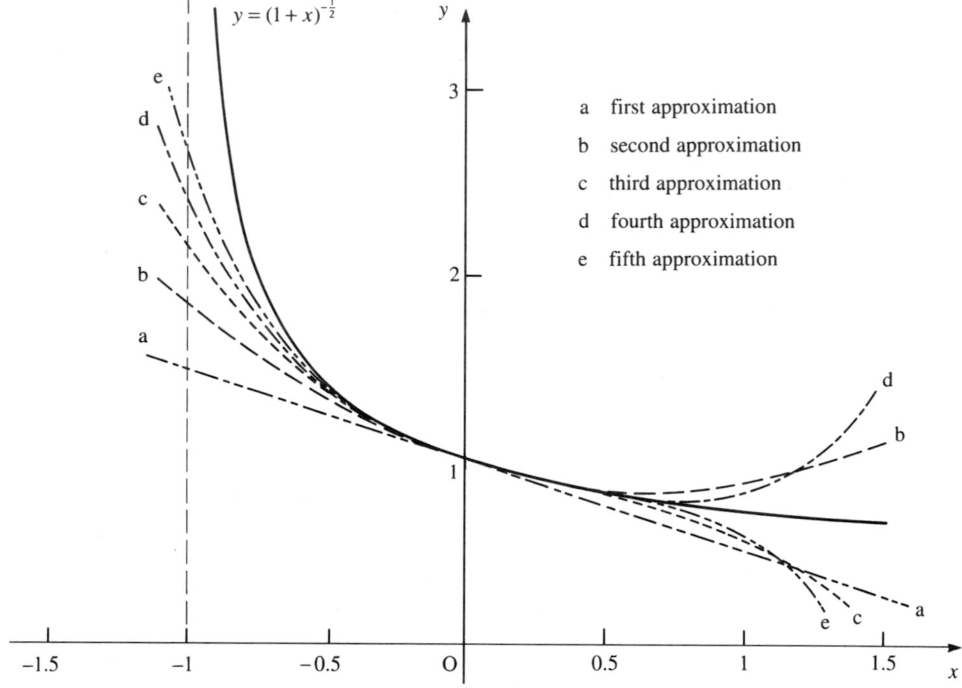

Figure 3.15

Activity

(i) Explain why it is not possible to find Maclaurin expansions for $\ln x$.

(ii) (a) Show that the Maclaurin series for $\ln(1 + x)$ is

$$x - \frac{x^2}{2} + \frac{x^3}{3} - \ldots + \frac{(-1)^{n+1}x^n}{n} + \ldots$$

(b) This series is valid for $-1 < x \leqslant 1$ only: by drawing graphs of $y = \ln(1 + x)$ and several successive approximations show that this is plausible.

(This series was first found by Nicolaus Mercator (1620–87), who lived for many years in London, though he was born in Denmark.)

Alternative approaches

Sometimes finding the coefficients of a Maclaurin series by repeated differentiation can be very laborious. As shown in the next example, the problem can be eased if you can express a derivative in terms of earlier derivatives, or the original function.

EXAMPLE

Find the first four non-zero terms of the Maclaurin series for $e^{2x}\sin 3x$.

Solution

Let $f(x) \equiv e^{2x}\sin 3x$. $\quad f(0) = 0$.

Then $f'(x) \equiv 2e^{2x}\sin 3x + 3e^{2x}\cos 3x$

$\equiv 2f(x) + 3e^{2x}\cos 3x \quad f'(0) = 3$.

and $f''(x) \equiv 2f'(x) + 6e^{2x}\cos 3x - 9e^{2x}\sin 3x$

$\equiv 2f'(x) + 2(f'(x) - 2f(x)) - 9f(x)$

$\equiv 4f'(x) - 13f(x). \quad f''(0) = 12$

Expressing $f''(x)$ in terms of $f'(x)$ and $f(x)$ simplifies further differentiation.

Then $f'''(x) \equiv 4f''(x) - 13f'(x) \quad f'''(0) = 48 - 39 = 9$

and $f^{(iv)}(x) \equiv 4f'''(x) - 13f''(x)$, etc. $\quad f^{(iv)}(0) = 36 - 156 = -120$

$$\text{Thus} \quad e^{2x}\sin 3x = 3x + \frac{x^2}{2!}\times 12 + \frac{x^3}{3!}\times 9 - \frac{x^4}{4!}\times 120 + \ldots$$

$$= 3x + 6x^2 + \frac{3}{2}x^3 - 5x^4 + \ldots.$$

Sometimes a Maclaurin series can be found by adapting one or more known Maclaurin series. Some such methods are indicated in the Activity below. You may well wonder whether the processes used are justifiable. Is it legitimate (for example) to integrate (or differentiate) an infinite series term by term? Can you form the product of two infinite series by multiplying terms? Is the series obtained identical to the series that would have been obtained by evaluating the derivatives? Answering these important questions in detail is beyond the scope of this book, though generally the answer is 'Yes, subject to certain conditions'.

Activity

Try out the following methods and explain why they work. How would you obtain further terms of the required series?

(i) The Maclaurin series for $\ln(1 + x)$ can be found by integrating the terms of the binomial series for $(1 + x)^{-1}$. Why is the integration constant zero?

(ii) The start of the Maclaurin series for $\dfrac{e^x}{1+x}$ can be found by multiplying together the first four terms of the Maclaurin series for e^x and $(1 + x)^{-1}$ and discarding all terms in x^4 and higher powers.

(iii) The first few terms of the Maclaurin series for $\sec x$ can be found from the first three terms of the Maclaurin series for $(1 + y)^{-1}$ where $y = -\frac{x^2}{2!} + \frac{x^4}{4!}$.

Taylor approximations

All Maclaurin expansions are 'centred' on $x = 0$. But it is possible to form expansions centred elsewhere:

let $g(h) \equiv f(a + h)$ where a is the constant $x - h$;

then $g'(h) \equiv f'(a + h)$, $g''(h) \equiv f''(a + h)$, etc.,

and $g(0) \equiv f(a)$, $g'(0) \equiv f'(a)$, $g''(0) \equiv f''(a)$, etc. so that

$$f(a + h) \equiv g(h) \approx g(0) + hg'(0) + \frac{h^2}{2!}g''(0) + \frac{h^3}{3!}g'''(0) + \ldots + \frac{h^n}{n!}g^{(n)}(0).$$

This may be expressed in either of the following two ways:

$$f(a + h) \approx f(a) + hf'(a) + \frac{h^2}{2!}f''(a) + \frac{h^3}{3!}f'''(a) + \ldots + \frac{h^n}{n!}f^{(n)}(a)$$

or equivalently

$$f(x) \approx f(a) + (x-a)f'(a) + \frac{(x-a)^2}{2!}f''(a) + \frac{(x-a)^3}{3!}f'''(a) + \ldots + \frac{(x-a)^n}{n!}f^{(n)}(a).$$

These two formulae are alternative versions of the nth *Taylor approximation* for $f(x)$ centred on $x = a$. They are also known as *Taylor polynomials*. (A Maclaurin approximation is a special case of a Taylor approximation, obtained by putting $a = 0$.)

Activity

Explain the connection between the first Taylor approximation for $f(x)$ and the Newton–Raphson method of approximating to the root of the equation $f(x) = 0$.

HISTORICAL NOTE

The Taylor approximations were discovered or rediscovered in various forms by several mathematicians in the seventeenth and eighteenth centuries. They were familiar to Scotsman James Gregory (1638–1675), though Englishman Brook Taylor (1685–1731) was the first to publish an account of them, in 1715. In 1742 Colin Maclaurin (1698–1745), Gregory's successor as professor at Edinburgh, published his expansion, stating that it occurred as a special case of Taylor's result; for some reason it has been credited to him as a separate theorem.

Exercise 3E

1 – 9. Find the Maclaurin series (up to the term in x^4) for each of the functions.

1. $\tan x$ **2.** $\sec x$

3. $\ln(1 + \sin x)$ **4.** $\sin 3x$

5. $\cos 2x$ **6.** $\arcsin x$

7. $\sin^2 x$ **8.** $e^{\sin x}$

9. e^{1+x}

10. A graphics calculator or graph-drawing software will be useful in this question.

(i) Draw a graph of $y = \sin x$. On the same axes draw graphs of the first few Maclaurin approximations to $\sin x$.

(ii) Repeat (i) for (a) $\cos x$; (b) $(1 + x)^{-1}$; (c) $(1 + x)^{0.5}$.

11. (i) Show that the Maclaurin series for $\arctan x$ is $x - \frac{x^3}{3} + \frac{x^5}{5} - \frac{x^7}{7} + \ldots$.

(This is known as Gregory's series, after the Scottish mathematician, James Gregory, who published it in 1668, well before Newton or Leibniz introduced calculus. The series is valid for $|x| \leq 1$.)

(ii) By putting $x = 1$ show that

$$\frac{\pi}{4} = 1 - \frac{1}{3} + \frac{1}{5} - \frac{1}{7} + \ldots.$$

(This is known as Leibniz's series. It converges very slowly.)

(iii) Show that (a) $\frac{\pi}{4} = \arctan\frac{1}{2} + \arctan\frac{1}{3}$

(known as Euler's formula for π),

(b) $\frac{\pi}{4} = 4\arctan\frac{1}{5} - \arctan\frac{1}{239}$

(known as Machin's formula).

(iv) Use Machin's formula together with Gregory's series to find the value of π to five decimal places. (In 1873 William Shanks used this method to calculate π to 707 decimal places, but he made a mistake in the 528th place, not discovered until 1946!)

12. Write down the Maclaurin series for $e^{-\frac{1}{2}x^2}$

(i) as far as x^6; (ii) as far as x^8.

It can be shown that $e^{-\frac{1}{2}x^2}$ always lies between these two approximations. Use them to estimate $\int_0^1 e^{-\frac{1}{2}x^2}\,dx$ and to establish error bounds for your answer. (It is not possible to find $\int e^{-\frac{1}{2}x^2}\,dx$ explicitly, but finding good approximations for integrals such as $\int_a^b e^{-\frac{1}{2}x^2}\,dx$ was an essential part of the construction of the normal distribution tables, a key tool in statistics.)

13. In this question give all numerical answers to four decimal places.

(i) Put $x = 1$ in the expansion

$$\ln(1 + x) \approx x - \frac{x^2}{2} + \frac{x^3}{3} - \ldots - \frac{x^{10}}{10}$$

and calculate an estimate of $\ln 2$. (Approximately 1000 terms would be needed to obtain $\ln 2$ correct to three decimal places by this method.)

(ii) Show that $\ln 2 = -\ln(1 - \frac{1}{2})$ and hence estimate $\ln 2$ by summing six terms.

(iii) Write down the series for $\ln(1 + x) - \ln(1 - x)$ as far as the first three non-zero terms and estimate ln 2 by summing these terms using a suitable value of x.

14. A curve passes through the point $(0, 2)$; its gradient is given by the differential equation $\frac{dy}{dx} = 1 - xy$. Assume that the equation of this curve can be expressed as the Maclaurin series

$$y = a_0 + a_1x + a_2x^2 + a_3x^3 + a_4x^4 + \ldots .$$

(i) Find a_0 and show that

$$a_1 + 2a_2x + 3a_3x^2 + 4a_4x^3 + \ldots$$
$$= 1 - 2x - a_1x^2 - a_2x^3 - a_3x^4 - \ldots .$$

(ii) Equate coefficients to find the first seven terms of the Maclaurin series.

(iii) Draw graphs to compare the solution given by these seven terms with a solution generated (step by step) on a computer.

Exercise 3E continued

15. Write down the term in x^r in the Maclaurin series for $\left(1+\frac{x}{k}\right)^k$ and show that it can be expressed as $\frac{(1-\frac{1}{k})(1-\frac{2}{k})(1-\frac{3}{k})\ldots(1-\frac{r-1}{k})}{r!}x^r$.

Use this result to explain why it is plausible that $\left(1+\frac{x}{k}\right)^k \to e^x$ as $k \to \infty$.

(This limit is an important part of the justification that under certain conditions a Poisson distribution can be used to approximate to a binomial distribution; see *Statistics 2*.)

16. (i) Prove by induction that

$$f(x) \equiv e^x \sin x \Rightarrow f^{(n)}(x) \equiv 2^{\frac{n}{2}} e^x \sin\left(x+\frac{n\pi}{4}\right)$$

and use this result to obtain the Maclaurin series for $e^x \sin x$ as far as x^6.

(ii) Multiply the third Maclaurin approximation for e^x by the third Maclaurin approximation for $\sin x$, and comment on what you notice.

(iii) Find a Maclaurin approximation for $e^x \cos x$ by multiplying the third Maclaurin approximation for e^x by the fourth Maclaurin approximation for $\cos x$, giving as many terms in your answer as you think justifiable.

17. (i) Show that no polynomials in x exactly represent (a) e^x, (b) $\sin x$, (c) $|x|$.

(ii) Show that $\ln(2e^x)$ can be represented exactly by a polynomial in x.

18. A projectile is launched from O with initial velocity $\begin{pmatrix} u \\ v \end{pmatrix}$ relative to horizontal and vertical axes through O. The path of the projectile may be modelled in various ways. The table below shows the position (x, y) of the projectile at time t after launch, as given by two different models. Both models assume that g (gravitational acceleration) is constant. Use Maclaurin expansions for e^{-kt}, where k is constant, to show that the results given by Model 1 are a special case of the results from Model 2, with $k = 0$.

	Assumptions about air resistance	**Position at time t**
Model 1	There is no air resistance.	$x = ut$ $y = vt - \frac{1}{2}gt^2$
Model 2	Air resistance is proportional to the velocity (with proportionality constant k).	$x = \frac{u}{k}\left(1-e^{-kt}\right)$ $y = \frac{g+kv}{k^2}\left(1-e^{-kt}\right)-\frac{gt}{k}$

Investigations

1. There are many ways of obtaining sequences of polynomial approximations for $f(x) \equiv \sin x$, for $0 \leqslant x \leqslant \frac{\pi}{2}$. Investigate alternative methods such as the following.

(i) Use (a) the linear function which passes through $(0, 0)$ and $(\frac{\pi}{2}, 1)$;

(b) the quadratic function which passes through (0, 0), $(\frac{\pi}{4}, \frac{1}{\sqrt{2}})$ and $(\frac{\pi}{2}, 1)$;

(c) the cubic function which passes through four points on $y = f(x)$; and so on.

(ii) Use polynomials $\mathrm{P}(x)$ which minimise

(a) $\int_0^{\frac{\pi}{2}} (\mathrm{f}(x) - \mathrm{P}(x))\,\mathrm{d}x$;

(b) the maximum value of $|\,\mathrm{f}(x) - \mathrm{P}(x)\,|$ in $0 \leqslant x \leqslant \frac{\pi}{2}$;

(c) $\int_0^{\frac{\pi}{2}} \left|\mathrm{f}(x) - \mathrm{P}(x)\right| \mathrm{d}x$;

(d) $\int_0^{\frac{\pi}{2}} (\mathrm{f}(x) - \mathrm{P}(x))^2\,\mathrm{d}x$.

2. What happens if (unconventionally) we decide to define $\arcsin x$ as:

(i) the angle y such that $x = \sin y$ where $\frac{\pi}{2} \leqslant y \leqslant \frac{3\pi}{2}$;

(ii) the angle y such that $x = \sin y$ where $\begin{cases} 0 \leqslant y \leqslant \frac{\pi}{2} & \text{if } 0 \leqslant x \leqslant 1; \\ \pi \leqslant y \leqslant \frac{3\pi}{2} & \text{if} \leqslant 1 \leqslant x < 0. \end{cases}$

KEY POINTS

- $\frac{\mathrm{d}}{\mathrm{d}x}\left(y^n\right) = ny^{n-1}\frac{\mathrm{d}y}{\mathrm{d}x}$
- $\frac{\mathrm{d}}{\mathrm{d}x}(\ln y) = \frac{1}{y}\frac{\mathrm{d}y}{\mathrm{d}x}$
- $\frac{\mathrm{d}}{\mathrm{d}x}(\mathrm{f}(x)y) = \mathrm{f}(x)\frac{\mathrm{d}y}{\mathrm{d}x} + \mathrm{f}'(x)y$
- $\frac{\mathrm{d}}{\mathrm{d}x}\left(y^n\frac{\mathrm{d}y}{\mathrm{d}x}\right) - y^n\frac{\mathrm{d}^2y}{\mathrm{d}x^2} + ny^{n-1}\left(\frac{\mathrm{d}y}{\mathrm{d}x}\right)^2$
- **Inverse trigonometric functions**

Function	Domain	Range	Derivative
$y = \arcsin x$	$-1 \leqslant x \leqslant 1$	$-\frac{\pi}{2} \leqslant y \leqslant \frac{\pi}{2}$	$\frac{1}{\sqrt{1-x^2}}$
$y = \arccos x$	$-1 \leqslant x \leqslant 1$	$0 \leqslant y \leqslant \pi$	$-\frac{1}{\sqrt{1-x^2}}$
$y = \arctan x$	all real x	$-\frac{\pi}{2} < y < \frac{\pi}{2}$	$\frac{1}{1+x^2}$

- $\int \frac{1}{a^2+x^2}\,dx = \frac{1}{a}\arctan\frac{x}{a} + c$ Use when integrating rational functions with constant numerator, and a quadratic denominator with no real roots.

- $\int \frac{1}{\sqrt{a^2-x^2}}\,dx = \arcsin\frac{x}{a} + c$ Use when integrating functions that can be arranged as a fraction, with constant numerator, and a denominator which is the square root of a quadratic; this quadratic must have distinct real roots, and the coefficient of x^2 must be negative.

- **Maclaurin series**

  General form:

$$f(x) = f(0) + x f'(0) + \frac{x^2}{2!}f''(0) + \frac{x^3}{3!}f'''(0) + \ldots + \frac{x^r}{r!}f^{(r)}(0) + \ldots$$

Valid for all x: $$e^x = 1 + x + \frac{x^2}{2!} + \frac{x^3}{3!} + \ldots + \frac{x^r}{r!} + \ldots$$

$$\sin x = x - \frac{x^3}{3!} + \frac{x^5}{5!} - \frac{x^7}{7!} + \ldots + \frac{(-1)^r x^{2r+1}}{(2r+1)!} + \ldots$$

$$\cos x = 1 - \frac{x^2}{2!} + \frac{x^4}{4!} - \frac{x^6}{6!} + \ldots + \frac{(-1)^r x^{2r}}{(2r)!} + \ldots$$

Valid for $|x| \leqslant 1$: $$\arctan x = x - \frac{x^3}{3} + \frac{x^5}{5} - \frac{x^7}{7} + \ldots + \frac{(-1)^{r-1} x^{2r-1}}{2r-1} + \ldots$$

Valid for $-1 < x \leqslant 1$: $$\ln(1+x) = x - \frac{x^2}{2} + \frac{x^3}{3} - \ldots + \frac{(-1)^{r-1} x^r}{r} + \ldots$$

Validity depends on n:

$$(1+x)^n = 1 + nx + \frac{n(n-1)}{2!}x^2 + \ldots + \frac{n(n-1)\ldots(n-r+1)}{r!}x^r + \ldots$$

If n is a positive integer: the series terminates after $n + 1$ terms, and is valid for all x.

If n is not a positive integer: the series is valid for $|x| < 1$; also for $|x| = 1$ if $n \geqslant -1$; and for $x = -1$ if $n > 0$.

4

Complex numbers

The shortest path between two truths in the real domain passes through the complex domain.

Jacques Hadamard, 1865–1963

How can complex numbers help in the design of the mechanism used to open the bin?

de Moivre's Theorem

First a reminder from *Pure Mathematics 4* (page 78): to multiply two complex numbers in polar form you *multiply* their moduli and *add* their arguments, so that, with the usual notation,

$$z_1z_2 = r_1r_2[\cos(\theta_1 + \theta_2) + j\sin(\theta_1 + \theta_2)].$$

Much can be achieved by using this repeatedly with just a single complex number z of unit modulus (so that we concentrate on what happens to the argument). For if $z = \cos\theta + j\sin\theta$

then

$$z^2 = \cos(\theta + \theta) + j\sin(\theta + \theta) = \cos 2\theta + j\sin 2\theta,$$

$$z^3 = z^2z = \cos(2\theta + \theta) + j\sin(2\theta + \theta) = \cos 3\theta + j\sin 3\theta,$$

and so on. This suggests the following general result.

de Moivre's Theorem

If n is any integer then

$$(\cos\theta + j\sin\theta)^n = \cos n\theta + j\sin n\theta.$$

Proof

The proof is in three parts, in which n is (i) positive, (ii) zero, or (iii) negative.

(i) When n is a positive integer we use induction.

The theorem is obviously true when $n = 1$, and if

$$(\cos\theta + j\sin\theta)^k = \cos k\theta + j\sin k\theta$$

then
$$\begin{aligned}(\cos\theta + j\sin\theta)^{k+1} &= (\cos k\theta + j\sin k\theta)(\cos\theta + j\sin\theta)\\ &= \cos(k\theta + \theta) + j\sin(k\theta + \theta)\\ &= \cos(k+1)\theta + j\sin(k+1)\theta\end{aligned}$$

so by induction the theorem is true for all positive integers n.

(ii) We define $z^0 = 1$ for all complex numbers $z \neq 0$. Therefore

$$(\cos\theta + j\sin\theta)^0 = 1 = \cos 0 + j\sin 0.$$

(iii) For negative n we deal first with the case $n = -1$. Since

$$(\cos\theta + j\sin\theta)(\cos(-\theta) + j\sin(-\theta)) = \cos(\theta - \theta) + j\sin(\theta - \theta) = 1$$

it follows that $(\cos\theta + j\sin\theta)^{-1} = \cos(-\theta) + j\sin(-\theta)$. ①

If n is a negative integer, let $n = -m$. Then

$$\begin{aligned}(\cos\theta + j\sin\theta)^n &= (\cos\theta + j\sin\theta)^{-m}\\ &= [(\cos\theta + j\sin\theta)^m]^{-1}\\ &= (\cos m\theta + j\sin m\theta)^{-1} \quad \text{using (i) for } m\text{, which is positive}\\ &= \cos(-m\theta) + j\sin(-m\theta) \quad \text{using ①}\\ &= \cos n\theta + j\sin n\theta.\end{aligned}$$

de Moivre's Theorem is also useful for simplifying powers of complex numbers when the modulus is not 1. For if $z = r(\cos\theta + j\sin\theta)$ then

$$z^n = [r(\cos\theta + j\sin\theta)]^n = r^n(\cos\theta + j\sin\theta)^n = r^n(\cos n\theta + j\sin n\theta).$$

EXAMPLE

Evaluate (i) $(\cos\pi/8 + j\sin\pi/8)^{12}$; (ii) $(\sqrt{3} + j)^5$.

Solution

(i) By de Moivre's Theorem

$$\begin{aligned}(\cos\pi/8 + j\sin\pi/8)^{12} &= \cos(12\times\pi/8) + j\sin(12\times\pi/8)\\ &= \cos(3\pi/2) + j\sin(3\pi/2)\\ &= -j.\end{aligned}$$

(ii) First convert to polar form:

$z = \sqrt{3} + j \Rightarrow |z| = \sqrt{(3+1)} = 2$, $\arg z = \arctan(1/\sqrt{3}) = \pi/6$.

So
$$\begin{aligned}(\sqrt{3} + j)^5 &= 2^5(\cos(\pi/6) + j\sin(\pi/6))^5\\ &= 32(\cos(5\pi/6) + j\sin(5\pi/6))\\ &= 32(-\sqrt{3}/2 + j/2)\\ &= -16\sqrt{3} + 16j.\end{aligned}$$

HISTORICAL NOTE

Abraham de Moivre (1667–1754) came to England from France as a Huguenot refugee at the age of eighteen and spent the rest of his long life in London. In papers from 1707 onwards he made use of 'his' theorem, though he never published it explicitly.

Exercise 4A

1 – 4. Use de Moivre's Theorem to evaluate the following.

1. $(\cos(\pi/4) + j\sin(\pi/4))^{15}$

2. $(\cos(\pi/3) + j\sin(\pi/3))^{-8}$

3. $(\cos(-\pi/12) + j\sin(-\pi/12))^{10}$

4. $(\cos(7\pi/8) - j\sin(7\pi/8))^{6}$

[**Hint:** $\cos\theta - j\sin\theta = \cos(-\theta) + j\sin(-\theta)$]

5 – 8. By converting to polar form and using de Moivre's Theorem, find the following in the form $x + jy$, giving x and y as exact expressions or correct to three decimal places.

5. $(1 - \sqrt{3}j)^4$

6. $(-2 + 2j)^7$

7. $(0.6 + 0.8j)^{-5}$

8. $(\sqrt{27} + 3j)^6$

9 – 12. Simplify the following.

9. $(\cos(-\alpha) + j\sin(-\alpha))^8$

10. $(\cos\beta + j\sin\beta)^3/(\cos\beta - j\sin\beta)^{-5}$

11. $(\cos^2\gamma + j\sin\gamma\cos\gamma)^{10}$

12. $(1 + \cos 2\delta + j\sin 2\delta)^{-4}$.

13. Deduce from de Moivre's Theorem that

$(\cos\theta - j\sin\theta)^n = \cos n\theta - j\sin n\theta$

(i) by putting $\theta = -\phi$;

(ii) by using conjugates.

Using de Moivre's Theorem

One of the reasons for the general acceptance of complex numbers during the eighteenth century was their usefulness in producing results involving only *real* numbers; these results could be obtained without using complex numbers, but often only with considerably greater trouble. De Moivre's Theorem is a good source of such examples.

EXAMPLE

Express $\cos 5\theta$ in terms of $\cos\theta$.

Solution

By de Moivre's Theorem

$$\begin{aligned}\cos 5\theta + j\sin 5\theta &= (\cos\theta + j\sin\theta)^5 \\ &= c^5 + 5jc^4s - 10c^3s^2 - 10jc^2s^3 + 5cs^4 + js^5\end{aligned}$$

(where the abbreviations c for $\cos\theta$ and s for $\sin\theta$ are used to save writing).

Equating real parts:

$$\cos 5\theta = c^5 - 10c^3s^2 + 5cs^4.$$

But $\quad s^2 = 1 - c^2$

so

$$\begin{aligned}\cos 5\theta &= c^5 - 10c^3(1 - c^2) + 5c(1 - c^2)^2 \\ &= c^5 - 10c^3 + 10c^5 + 5c - 10c^3 + 5c^5.\end{aligned}$$

Therefore $\cos 5\theta = 16\cos^5\theta - 20\cos^3\theta + 5\cos\theta$.

Activity

(i) Check that this gives the correct results when $\theta = 0$ and when $\theta = \pi$.

(ii) By equating imaginary parts find $\sin 5\theta$ in terms of $\sin\theta$.

Notice that de Moivre not only gives a straightforward solution of the original problem, but also gives the expression for $\sin 5\theta$ with very little extra work – two for the price of one! Sometimes it is worth setting out to do *more* than is required, as in the next example.

EXAMPLE

Find a simplified expression for the sum of the series

$$1 + {}^nC_1\cos\theta + {}^nC_2\cos 2\theta + {}^nC_3\cos 3\theta + \ldots + \cos n\theta.$$

Solution

At first sight this series suggests the binomial expansion $(1 + \cos\theta)^n$: the coefficients $1\ (= {}^nC_0)$, ${}^nC_1$, ${}^nC_2$, ..., $1\ (= {}^nC_n)$ are right, but there are multiple angles, $\cos r\theta$, instead of powers of cosines, $\cos^r\theta$. This indicates that de Moivre's Theorem can be used, and so it is worth introducing the corresponding sine series too.

Let $\quad C = 1 + {}^nC_1\cos\theta + {}^nC_2\cos 2\theta + {}^nC_3\cos 3\theta + \ldots + \cos n\theta$

and $\quad S = {}^nC_1\sin\theta + {}^nC_2\sin 2\theta + {}^nC_3\sin 3\theta + \ldots + \sin n\theta.$

Then

$$\begin{aligned} C + jS &= 1 + {}^nC_1(\cos\theta + j\sin\theta) + {}^nC_2(\cos 2\theta + j\sin 2\theta) + \ldots + (\cos n\theta + j\sin n\theta) \\ &= 1 + {}^nC_1 z + {}^nC_2 z^2 + \ldots + z^n \quad (\text{where } z = \cos\theta + j\sin\theta) \\ &= (1 + z)^n. \end{aligned}$$

This expresses $C + jS$ compactly, but we still have to separate the real and imaginary parts. To do this we use the double angle formulae, but using $\theta/2$ rather than θ:

$$\cos\theta = 2\cos^2\tfrac{\theta}{2} - 1 \text{ and } \sin\theta = 2\sin\tfrac{\theta}{2}\cos\tfrac{\theta}{2}.$$

So $1 + z = 1 + \cos\theta + j\sin\theta$

$$\begin{aligned} &= 2\cos^2\tfrac{\theta}{2} + 2j\sin\tfrac{\theta}{2}\cos\tfrac{\theta}{2} \\ &= 2\cos\tfrac{\theta}{2}\left(\cos\tfrac{\theta}{2} + j\sin\tfrac{\theta}{2}\right) \end{aligned}$$

Such half-angle manipulations often occur in this sort of work.

Therefore $\quad (1 + z)^n = 2^n\cos^n\tfrac{\theta}{2}\left(\cos\tfrac{\theta}{2} + j\sin\tfrac{\theta}{2}\right)^n$

$$= 2^n\cos^n\tfrac{\theta}{2}\left(\cos\tfrac{n\theta}{2} + j\sin\tfrac{n\theta}{2}\right).$$

Taking the real part, $C = 2^n\cos^n\tfrac{\theta}{2}\cos\tfrac{n\theta}{2}$.

Activity

State the result obtained by equating imaginary parts.

The example on page 71 gave a multiple-angle formula in terms of powers; it is sometimes useful (for example when integrating) to do the reverse. For this we need the following deduction from the main theorem:

if $\quad z = \cos\theta + j\sin\theta$

then $\quad z^n = \cos n\theta + j\sin n\theta$

and $\quad z^{-n} = \cos(-n\theta) + j\sin(-n\theta) = \cos n\theta - j\sin n\theta.$

Therefore $\quad \cos n\theta = (z^n + z^{-n})/2$

and $\quad \sin n\theta = (z^n - z^{-n})/2j.$

EXAMPLE

Express $\cos^5\theta$ in terms of multiple angles.

Solution

Let $\quad z = \cos\theta + j\sin\theta$. Then

$$2\cos\theta = z + z^{-1}$$

$$\Rightarrow \quad 2^5\cos^5\theta = (z + z^{-1})^5$$

$$= z^5 + 5z^3 + 10z + 10z^{-1} + 5z^{-3} + z^{-5}$$

$$= (z^5 + z^{-5}) + 5(z^3 + z^{-3}) + 10(z + z^{-1})$$

$$= 2\cos 5\theta + 10\cos 3\theta + 20\cos\theta$$

$$\Rightarrow \quad \cos^5\theta = (\cos 5\theta + 5\cos 3\theta + 10\cos\theta)/16$$

Activity

Use a similar method to express $\sin^5\theta$ in terms of multiple angles.

Exercise 4B

1. Prove that $\cos 4\theta = c^4 - 6c^2s^2 + s^4$ and $\sin 4\theta = 4c^3s - 4cs^3$, where $c = \cos\theta$, $s = \sin\theta$.

Use these results to find $\tan 4\theta$ as a rational function of $\tan\theta$.

[**Hint:** put $\tan 4\theta = \dfrac{\sin 4\theta}{\cos 4\theta}$ and divide throughout by c^4.]

2. Find the expressions for $\cos 3\theta$ and $\sin 3\theta$ given by de Moivre's Theorem. Hence express

(i) $\cos 3\theta$ in terms of $\cos\theta$;

(ii) $\sin 3\theta$ in terms of $\sin\theta$;

(iii) $\tan 3\theta$ in terms of $\tan\theta$.

3. Find $\cos 6\theta$ and $\sin 6\theta/\sin\theta$ in terms of $\cos\theta$.

4. If $c = \cos\theta$, $s = \sin\theta$, $t = \tan\theta$ show that

$$\cos n\theta = c^n - {}^nC_2c^{n-2}s^2 + {}^nC_4c^{n-4}s^4 - \ldots$$
$$= c^n(1 - {}^nC_2t^2 + {}^nC_4t^4 - \ldots)$$

and
$$\sin n\theta = {}^nC_1c^{n-1}s - {}^nC_3c^{n-3}s^3 + \ldots$$
$$= c^n({}^nC_1t - {}^nC_3t^3 + \ldots).$$

Hence find $\tan n\theta$ in terms of t.

5 – 9. Express each of the following in terms of multiple angles.

5. $\cos^4\theta$

6. $\sin^5\theta$

7. $\sin^6\theta$

8. $\cos^3\theta\sin^4\theta$

9. $\cos^4\theta\sin^3\theta$

Exercise 4B continued

10. Prove that $\cos^m\theta \sin^n\theta$ can be expressed in terms of the cosines of multiple angles if n is even, and in terms of the sines of multiple angles if n is odd.

11 – 13. Use your previous results to find these integrals.

11. $\int \sin^6\theta \, d\theta$

12. $\int^{\pi/2} \cos^3\theta \sin^4\theta \, d\theta$

13. $\int_0^{\pi} \cos^4\theta \sin^3\theta \, d\theta$

14. Use $\cos n\theta = (z^n + z^{-n})/2$ to express

$\cos\theta + \cos 3\theta + \cos 5\theta + \ldots + \cos(2n-1)\theta$

as a geometric series in terms of z. Hence find this sum in terms of θ.

15. Let $C = 1 + \cos\theta + \cos 2\theta + \ldots + \cos(n-1)\theta$

and $S = \sin\theta + \sin 2\theta + \ldots + \sin(n-1)\theta$.

Show that $C + jS$ is a geometric progression with common ratio $z = \cos\theta + j\sin\theta$.

Hence show that

$C = \dfrac{1 - \cos\theta + \cos(n-1)\theta - \cos n\theta}{2 - 2\cos\theta}$, and find S.

16. (i) Show the points 2 and $2 + \cos(2\pi/3) + j\sin(2\pi/3)$ on an Argand diagram, and hence express $2 + \cos(2\pi/3) + j\sin(2\pi/3)$ in polar form.

(ii) Deduce that

$$\sum_{r=0}^{n} {}^nC_r 2^{n-r} \cos(2r\pi/3) = 3^{n/2} \cos(n\pi/6)$$

(iii) State the corresponding result for sines.

17. Sum the series

$$\sum_{r=0}^{n} {}^nC_r \sin(\alpha + r\beta).$$

18. By expressing $\cos^{2n}\theta$ in terms of cosines of multiple angles, prove that

$$\int_0^{\pi} \cos^{2n}\theta \, d\theta = \frac{(2n)!\pi}{2^{2n}(n!)^2}.$$

What is $\int_0^{\pi} \cos^{2n+1}\theta \, d\theta$?

Complex roots: the roots of unity

As early as 1629 Albert Girard stated that every polynomial equation of degree n has exactly n roots (including repetitions); this was first proved by the 18 year old Carl Friedrich Gauss 170 years later.

Therefore even the simple equation $z^n = 1$ has n roots. Of course one of these is $z = 1$, and if n is even then $z = -1$ is another. But where are the rest?

Activity

(i) Write down the two roots of $z^2 = 1$, and show them in an Argand diagram.

(ii) Use $z^3 - 1 = (z-1)(z^2 + z + 1)$ to find the three roots of $z^3 = 1$. Show them in an Argand diagram.

(iii) Find the four roots of $z^4 = 1$, and show them in an Argand diagram.

Every root of the equation $z^n = 1$ must have unit modulus, since otherwise the modulus of z^n would not be 1. So every root is of the form $z = \cos\theta + j\sin\theta$,

and $z^n = 1 \Leftrightarrow (\cos\theta + j\sin\theta)^n = 1$

$\Leftrightarrow \cos n\theta + j\sin n\theta = 1$ (by de Moivre)

$\Leftrightarrow n\theta = 2k\pi$, where k is any integer, since 1 is $(1, 0)$ or $(1, 2\pi)$ or $(1, 4\pi)$ or ... in polar form.

As k takes the values $0, 1, 2, ..., n-1$ the corresponding values of θ are 0, $\frac{2\pi}{n}, \frac{4\pi}{n}, ..., \frac{2(n-1)\pi}{n}$, giving n distinct values of z. But when $k = n$ then $\theta = 2\pi$, which gives the same z as $\theta = 0$. Similarly any integer value of k larger than n differs from one of $0, 1, 2, ..., n-1$ by a multiple of n, and so gives a value of θ differing by a multiple of 2π from one already listed; the same applies when k is any negative integer.

Therefore the equation $z^n = 1$ has precisely n roots. These are

$$z = \cos\frac{2k\pi}{n} + j\sin\frac{2k\pi}{n}, \quad k = 0, 1, 2, \ldots, n-1.$$

These n complex numbers are called the nth *roots of unity*. They include $z = 1$ when $k = 0$ and, if n is even, $z = -1$ when $k = n/2$. It is customary to use ω (the Greek letter omega) for the root with the smallest positive argument:

$$\omega = \cos\frac{2\pi}{n} + j\sin\frac{2\pi}{n}.$$

Then by de Moivre's Theorem $\omega^k = \cos\frac{2k\pi}{n} + j\sin\frac{2k\pi}{n}$, so that the nth roots of unity may be written as $1, \omega, \omega^2, \ldots, \omega^{n-1}$.

The complex numbers $1, \omega, \omega^2, \ldots, \omega^{n-1}$ are represented on the Argand diagram by the vertices of a regular n-sided polygon inscribed in the unit circle with one vertex at the point 1.

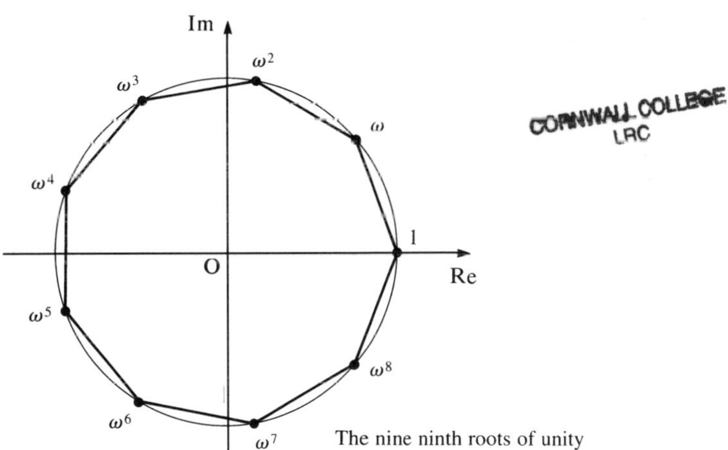

Figure 4.1

Activity

Prove that $(\omega^r)^* = \omega^{n-r}$.

The sum of all the nth roots of unity is a geometric series with common ratio ω:

$$1 + \omega + \omega^2 + \ldots + \omega^{n-1} = (1 - \omega^n)/(1 - \omega)$$
$$= 0$$

since $\omega^n = 1$

Therefore the sum of all n of the nth roots of unity is zero.

EXAMPLE Solve the equation $(1 + jz)^n = (1 - jz)^n$, where n is odd.

Solution

The equation can be rearranged as $\left(\dfrac{1+jz}{1-jz}\right)^n = 1$.

By taking the nth root of both sides we have

$\dfrac{1+jz}{1-jz} = \alpha$, where $\alpha = \cos\theta + j\sin\theta$ is an nth root of unity.

Solving this for z gives $z = \dfrac{\alpha - 1}{j(\alpha + 1)}$, where since n is odd $\alpha + 1 \neq 0$.

But $\quad \alpha - 1 = \cos\theta - 1 + j\sin\theta = -2\sin^2\frac{\theta}{2} + 2j\sin\frac{\theta}{2}\cos\frac{\theta}{2}$

$= 2j\sin\frac{\theta}{2}(\cos\frac{\theta}{2} + j\sin\frac{\theta}{2})$

and $\quad \alpha + 1 = \cos\theta + 1 + j\sin\theta = 2\cos^2\frac{\theta}{2} + 2j\sin\frac{\theta}{2}\cos\frac{\theta}{2}$

$= 2\cos\frac{\theta}{2}(\cos\frac{\theta}{2} + j\sin\frac{\theta}{2})$.

Substituting these in the expression for z and simplifying gives

$$z = \sin\tfrac{\theta}{2} / \cos\tfrac{\theta}{2} = \tan\tfrac{\theta}{2}.$$

So the roots are $z = \tan\dfrac{k\pi}{n}$, $k = 0, 1, 2, \ldots, n - 1$.

Activity

Work through the same example in the case when n is even. (Be careful: what is the degree of the equation now?)

Exercise 4C

1. Explain geometrically why the set of tenth roots of unity is the same as the set of fifth roots of unity together with their negatives.

2. If ω is a complex cube root of unity, $\omega \neq 1$, prove that

(i) $(1 + \omega)(1 + \omega^2) = 1$;

(ii) $1 + \omega$ and $1 + \omega^2$ are complex cube roots of -1;

(iii) $(a + b)(a + \omega b)(a + \omega^2 b) = a^3 + b^3$;

(iv) $(a + b + c)(a + \omega b + \omega^2 c)(a + \omega^2 b + \omega c)$
$= a^3 + b^3 + c^3 - 3abc$.

3. A regular hexagon is inscribed in the unit circle. One vertex is α. Give the other vertices in terms of α and ω, where ω is a complex cube root of unity.

4. The complex numbers $1, \omega, \omega^2, \ldots, \omega^{n-1}$ are represented as *vectors* in the Argand diagram, following 'nose to tail' in order. Explain geometrically why these form a regular polygon. Hence prove again that the sum of all the nth roots of unity is zero.

5. (i) (a) Draw an Argand diagram showing the points $1, \omega, \omega^2, \omega^3, \omega^4$, where $\omega = \cos(2\pi/5) + j\sin(2\pi/5)$.

(b) If $\alpha = \omega^2$ show that the points $1, \alpha, \alpha^2, \alpha^3, \alpha^4$ are the same as the points in (a), but in a different order. Indicate this order by joining successive points on your diagram.

(c) Repeat (b) with α replaced by β, where $\beta = \omega^3$.

(ii) Repeat the whole of (i) taking $\omega = \cos(2\pi/6) + j\sin(2\pi/6)$ and considering the points $1, \omega, \ldots, \omega^5$; $1, \alpha, \ldots, \alpha^5$; $1, \beta, \ldots, \beta^5$.

(iii) Do likewise for the seventh and eighth roots of unity.

(iv) If $\omega = \cos(2\pi/n) + j\sin(2\pi/n)$ and $\alpha = \omega^m$, form a conjecture about when

$$\{1, \omega, \omega^2, ..., \omega^{n-1}\} = \{1, \alpha, \alpha^2, ..., \alpha^{n-1}\}.$$

6. Solve the equation $z^3 = (j - z)^3$.

7. Solve the equation $z^5 + z^4 + z^3 + z^2 + z + 1 = 0$.

8. Prove that all the roots of $(z - 1)^n = z^n$ have real part 1/2.

9. Solve the equation $(j - z)^n = (jz - 1)^n$.

10. Solve the equation $(z + j)^n + (z - j)^n = 0$.

Complex roots: the general case

To find the nth roots of any given non-zero complex number w you have to find z such that $z^n = w$. The pattern of argument is the same as in the previous section on nth roots of unity, but adjusted to take account of the modulus s and argument ϕ of w. So let

$$z = r(\cos\theta + j\sin\theta) \text{ and } w = s(\cos\phi + j\sin\phi).$$

$$\text{Then } z^n = w \Leftrightarrow r^n(\cos\theta + j\sin\theta)^n = s(\cos\phi + j\sin\phi)$$
$$\Leftrightarrow r^n(\cos n\theta + j\sin n\theta) = s(\cos\phi + j\sin\phi).$$

Two complex numbers in polar form are equal only if they have the same moduli and their arguments are equal or differ by a multiple of 2π. Therefore

$$r^n = s \text{ and } n\theta = \phi + 2k\pi, \text{ where } k \text{ is an integer.}$$

Since r and s are positive real numbers the equation $r^n = s$ gives the *unique* value $r = s^{1/n}$, so all the roots lie on the circle $|z| = s^{1/n}$.

The argument of z is $\theta = \dfrac{\phi + 2k\pi}{n}$. As k take the values 0, 1, 2, ..., $n - 1$, this gives n distinct complex numbers z, and (by the same argument as for the roots of unity) there are no others.

Therefore the non-zero complex number $s(\cos\phi + j\sin\phi)$ has precisely n different nth roots. These are

$$s^{1/n}\left(\cos\left(\frac{\phi + 2k\pi}{n}\right) + j\sin\left(\frac{\phi + 2k\pi}{n}\right)\right) \quad \text{where } k = 0, 1, 2, ..., n - 1.$$

EXAMPLE

Represent $2 - 2j$ and its five fifth roots on an Argand diagram.

Solution

Since $2 - 2j = 8^{1/2}(\cos(-\pi/4) + j\sin(-\pi/4))$, the fifth roots all have modulus $8^{1/10} \approx 1.23$.

Their arguments are

$$-\pi/20, -\pi/20 + 2\pi/5, -\pi/20 + 4\pi/5, -\pi/20 + 6\pi/5, -\pi/20 + 8\pi/5$$

or (taking principal arguments in degrees) –9°, 63°, 135°, –153°, –81°.

The fifth roots are the vertices of a regular pentagon inscribed in the circle $|z| = 8^{1/10}$, figure 4.2.

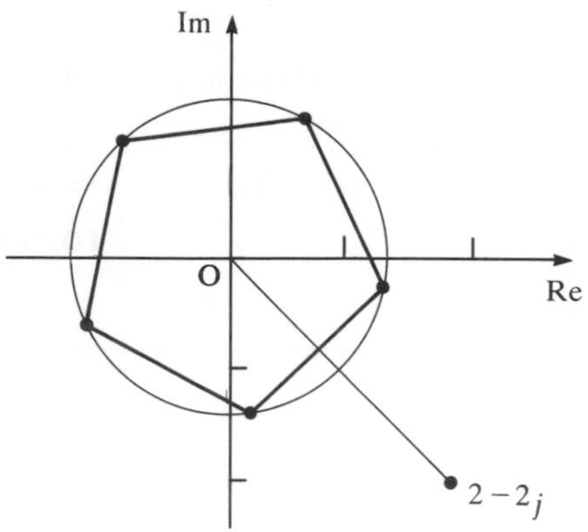

Figure 4.2

Activity

Express $8^{1/10}(\cos(-\pi/20) + j\sin(-\pi/20))$ in the form $x + yj$, giving x and y correct to two decimal places.

Figure 4.2 is typical of the general case: the n nth roots of z are represented by the vertices of a regular n-gon inscribed in the circle with centre O and radius $|z|^{1/n}$. This can be useful when dealing with the geometry of regular polygons.

EXAMPLE

The vertices $A_0, A_1, \ldots, A_{n-1}$ of a regular n-gon lie on a circle of unit radius with centre O. The point P is such that $\overrightarrow{OP} = 3\overrightarrow{OA_0}$. Prove that

$$(PA_0)^2 + (PA_1)^2 + \ldots + (PA_{n-1})^2 = 10n.$$

Solution

Let $\omega = \cos(2\pi/n) + j\sin(2\pi/n)$, an nth root of unity. Then the vertices A_r represent the complex numbers ω^r for $r = 0, 1, \ldots, n-1$, and P represents 3. Therefore

$$\begin{aligned}(PA_r)^2 &= |\omega^r - 3|^2 = (\omega^r - 3)(\omega^r - 3)^* = (\omega^r - 3)(\omega^{n-r} - 3)\\ &= \omega^n - 3\omega^r - 3\omega^{n-r} + 9\\ &= 10 - 3\omega^r - 3\omega^{n-r}, \text{ since } \omega^n = 1.\end{aligned}$$

When this expression is summed from $r = 0$ to $r = n - 1$ the first term gives $10n$ and each of the two sums involving ω is zero, since $1 + \omega + \omega^2 + \ldots + \omega^{n-1} = 0$. This proves the required result.

If $w = s(\cos\phi + j\sin\phi)$ then $w^m = s^m(\cos m\phi + j\sin m\phi)$ for all integers m by de Moivre's Theorem.

The complex number w^m has the n nth roots

$$s^{m/n}\left(\cos\left(\frac{m\phi + 2k\pi}{n}\right) + j\sin\left(\frac{m\phi + 2k\pi}{n}\right)\right).$$

One of these is $s^{m/n}\left(\cos\left(\frac{m\phi}{n}\right)+j\sin\left(\frac{m\phi}{n}\right)\right)$, and we use the notation

$w^{m/n}$ to mean *this* nth root of w^m. This definition ensures that de Moivre's Theorem is also true for rational powers, since

$$(\cos\theta + j\sin\theta)^{m/n} = s^{m/n}\left(\cos\left(\frac{m\theta}{n}\right)+j\sin\left(\frac{m\theta}{n}\right)\right).$$

For Discussion

Explain the fallacy in the following argument:

'$j = \sqrt{(-1)} = \sqrt{(1/-1)} = \sqrt{1}/\sqrt{(-1)} = 1/j$, so $j^2 = 1$.

But $j^2 = -1$. Therefore $1 = -1$.'

Exercise 4D

1. Find both square roots of $-7 + 5j$, giving your answers in the form $x + yj$ with x and y correct to two decimal places.

2. Find the four fourth roots of -4, giving your answers in the form $x + yj$, and show them on the Argand diagram.

3. One fourth root of w is $2 + 3j$. Find w and its other fourth roots, and represent all five points on the Argand diagram.

4. Represent the five solutions of the equation $(z - 3j)^5 = 32$ on the Argand diagram.

5. A regular heptagon (seven sides) on the Argand diagram has centre $-1 + 3j$ and one vertex at $2 + 3j$. Write down the equation whose solutions are represented by the vertices of this heptagon.

6. One of the nth roots of w is α. Prove that the other roots are $\alpha\omega, \alpha\omega^2, \ldots, \alpha\omega^{n-1}$ where $w = \cos(2\pi/n) + j\sin(2\pi/n)$. Deduce that the sum of all the nth roots of w is zero.

7. The nth roots of w are represented by vectors on the Argand diagram, with $w^{1/n}$ as a position vector and with each subsequent vector added to its predecessor. Describe the figure which is formed, and deduce again that the sum of all the nth roots is zero.

8. The vertices A_1, A_2, A_3, A_4, A_5 of a regular pentagon lie on a circle of unit radius with centre at the point O. A_1 is the midpoint of OP. Prove that

(i) $PA_1 \times PA_2 \times PA_3 \times PA_4 \times PA_5 = 31$;

(ii) $\sum_{n=1}^{5} (PA_n)^2 = 25$;

(iii) $A_1A_2 \times A_1A_3 \times A_1A_4 \times A_1A_5 = 5$.

9. There are n points equally spaced around the circumference of a circle of radius a. Prove that the sum of the squares of their distances from any diameter is $na^2/2$. Find the sum of the squares of their distances from any tangent.

10. (i) By considering the solutions of the equation $z^n - 1 = 0$ prove that

$$(z-\omega)(z-\omega^2)(z-\omega^3)\ldots(z-\omega^{n-1}) = z^{n-1} + z^{n-2} + \ldots + z + 1,$$

where $\omega = \cos(2\pi/n) + j\sin(2\pi/n)$.

(ii) There are n points equally spaced around the circumference of a unit circle. Prove that the product of the distances from one of these points to each of the others is n. (Question 8 part (iii) is the case $n = 5$.)

(iii) By finding expressions for the distances in (ii), deduce that

$$\sin\frac{\pi}{n}\sin\frac{2\pi}{n}\sin\frac{3\pi}{n}\ldots\sin\frac{(n-1)\pi}{n} = \frac{n}{2^{n-1}}.$$

11 – 13. Find the following in polar form.

11. $[32(\cos(\pi/6) + j\sin(\pi/6))]^{3/5}$

12. $[343(\cos(3\pi/8) + j\sin(3\pi/8))]^{-2/3}$

Exercise 4D continued

13. $[81(\cos(-\pi/3) + j\sin(-\pi/3))]^{-1.75}$

14. (i) Find $(j^{1/2})^3$ and $(j^3)^{1/2}$. Which of these is $j^{3/2}$?

(ii) Find $(j^{1/3})^2$ and $(j^2)^{1/3}$.

(iii) Find a condition involving m and $\arg w$ which ensures that $(w^{1/n})^m = (w^m)^{1/n}$.

15. The Polish mathematician Hoëné Wronski (1778–1853) once wrote that

$$\pi = \frac{2\infty}{\sqrt{-1}}\left\{(1+\sqrt{-1})^{1/\infty} - (1-\sqrt{-1})^{1/\infty}\right\}.$$

Was Wronski wrong?

Geometrical uses of complex numbers

Your study of complex numbers started in *Pure Mathematics 4* with their origin in algebra, in connection with the solution of polynomial equations. Then the simple idea of representing a complex number as a point or a vector in the Argand diagram soon made it possible for you to use complex numbers in geometry too. Some of these geometrical applications, such as the use of midpoints, other points of subdivision, centroids, and enlargements, can be handled equally well by two-dimensional vector methods. But with other problems, especially those involving rotations or similarity, complex number methods are especially effective. For example, you have seen in the previous section some fruitful links between regular polygons and the complex roots of unity.

Much of the geometrical power of complex numbers comes from the crucial result about the multiplication of complex numbers in polar form: 'multiply the moduli, add the arguments'. This means that the effect of multiplying a complex number z by a complex number λ is to turn the vector z in the Argand diagram through the angle $\arg\lambda$ anticlockwise and stretch it by the scale factor $|\lambda|$ to give the vector λz.

In particular, if A, B, C represent the numbers a, b, c in the Argand diagram, then $\overrightarrow{BA}$ represents the complex number $a - b$, while $\overrightarrow{BC}$ represents $c - b$ [figure 4.3]. Therefore

$$\text{if} \quad \frac{a-b}{c-b} = \lambda \quad \text{then} \quad \text{angle ABC} = \arg\lambda = \arg\left(\frac{a-b}{c-b}\right).$$

Angle ABC here means the *anticlockwise* angle through which $\overrightarrow{BC}$ has to be turned to bring it into line with $\overrightarrow{BA}$.

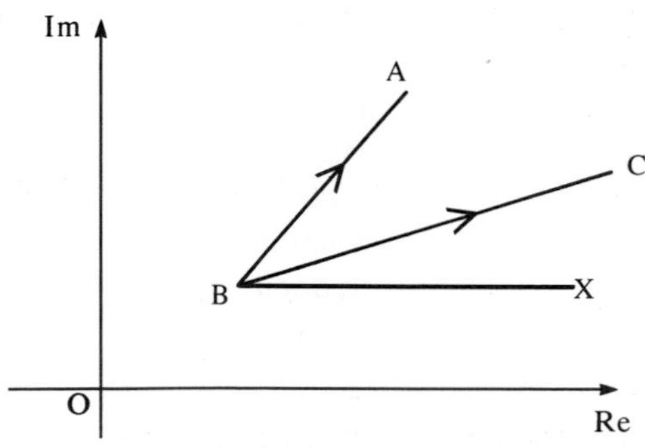

Figure 4.3

Activity

Prove the same result by starting with $\arg\left(\frac{a-b}{c-b}\right) = \arg(a-b) - \arg(c-b)$.

EXAMPLE

Find the locus of points z for which $\arg\left(\frac{z-2j}{z+3}\right) = \pi/3$.

Solution

Let A, B, P be the points representing $2j$, -3, z respectively.

The given condition shows that the direction of $\overrightarrow{AP}$ $(= z - 2j)$ is $\pi/3$ ahead of the direction of $\overrightarrow{BP}$ $(= z + 3)$, in the anticlockwise sense. Therefore $\angle APB = \pi/3$, and so (using the converse of the 'angles in the same segment' circle property) P lies on the arc of the circle with end points A and B as shown.

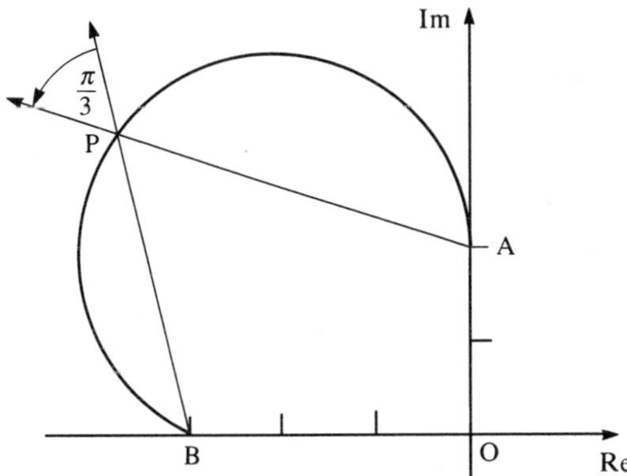

Activity

Find in a similar form the condition for P to lie on the other arc of this circle.

EXAMPLE

Two similar figures are *directly* similar if corresponding points moving round the two figures go in the same sense, either both clockwise or both anti-clockwise. Find a condition for two triangles in the Argand diagram to be directly similar.

Solution

Let A, B, C, D, E, F be the points representing a, b, c, d, e, f (see the diagram overleaf).

Then triangles ABC, DEF are directly similar if and only if

$$\frac{AB}{BC} = \frac{DE}{EF} \quad \text{and} \quad \text{angle ABC} = \text{angle DEF},$$

both angles being in the same sense because the similarity is direct.

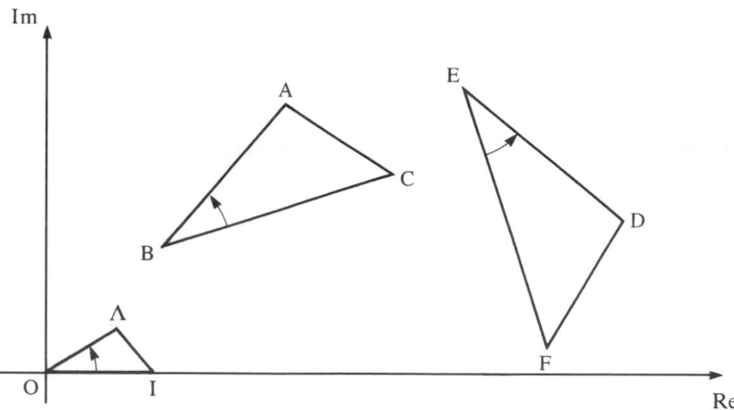

Therefore $\dfrac{|a-b|}{|c-b|}=\dfrac{|d-e|}{|f-e|}$ and $\arg\left(\dfrac{a-b}{c-b}\right)=\arg\left(\dfrac{d-e}{f-e}\right)$,

and so $\dfrac{a-b}{c-b}=\dfrac{d-e}{f-e}$.

NOTE

The shape of these triangles is determined by the single complex number $\lambda=\dfrac{a-b}{c-b}=\dfrac{d-e}{f-e}$, *and both triangles are similar to the triangle ΛOI with vertices* λ, 0, 1 *(see the diagram above). When dealing with a set of similar triangles it can be helpful to make use of ΛOI as the 'standard representative' of the whole family of triangles with this shape.*

Activity

Prove that the condition for two triangles ABC, DEF to have opposite similarity (where corresponding points move in opposite senses round the two triangles) is

$$\frac{a-b}{c-b}=\left(\frac{d-e}{f-e}\right)^{*}.$$

EXAMPLE

Squares whose centres are P, Q, R are drawn outwards on the sides BC, CA, AB respectively of a triangle ABC. Prove that AP and QR are equal and mutually perpendicular.

Solution

Working on an Argand diagram, let the points A, B, ... correspond to the complex numbers a, b, ... as usual (see the diagram opposite).

The first step is to find p in terms of b and c: two ways of doing this are given:

(i) If the vertex opposite B of the square with centre P is D then $\overrightarrow{CD}$ is obtained by turning $\overrightarrow{CB}$ through a right angle anticlockwise, and so $d-c=j(b-c)$. Therefore, since P is the midpoint of BD,

$$p=\frac{b+d}{2}=\frac{b+c+j(b-c)}{2}.$$

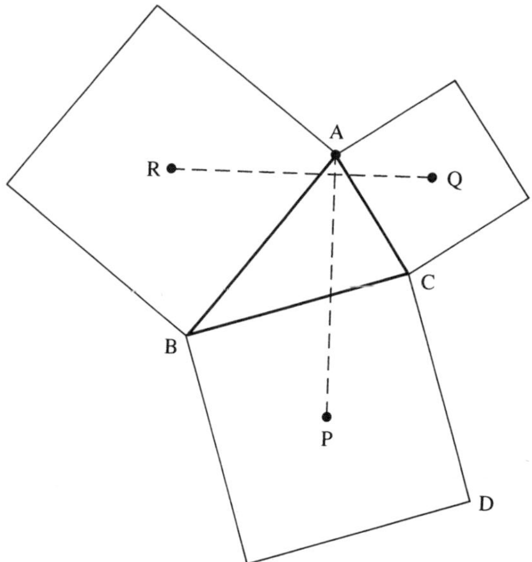

(ii) Alternatively, triangles BCP, CAQ, ABR are all right angled isosceles triangles, and therefore all similar to the 'standard representative' triangle AOI with vertices $\frac{1+j}{2}$, 0, 1.

Hence $\quad \frac{p-c}{b-c} = \frac{q-a}{c-a} = \frac{r-b}{a-b} = \frac{1+j}{2}$, from which

$p = c + \frac{(1+j)(b-c)}{2} = \frac{b+c+j(b-c)}{2}$, as before, and similarly

$q = \frac{c+a+j(c-a)}{2}$ and $r = \frac{a+b+j(a-b)}{2}$.

It is now easy to complete the proof:

$$\begin{aligned} 2(q-r) &= c + a + j(c-a) - a - b - j(a-b) \\ &= c - b + j(b+c) - 2ja \\ &= j(b + c + j(b-c) - 2a) \\ &= 2j(p-a). \end{aligned}$$

Therefore $\overrightarrow{RQ}$ is obtained by turning $\overrightarrow{AP}$ through a right angle.

Exercise 4E

1. Explain the following construction for multiplication of given numbers z_1 and z_2 in the Argand diagram:

Draw the triangle with vertices 0, 1, z_1. Then construct the directly similar triangle which has vertices 0 and z_2 corresponding to vertices 0 and 1 of the original triangle. The third vertex of the constructed triangle is $z_1 z_2$.

Illustrate this by doing the construction for $z_1 = 2 + j$, $z_2 = 3 + 4j$. Then do the same with z_1 and z_2 interchanged.

2. Find the locus of points z for which $\arg\left(\frac{z-2j}{z+3}\right) = -\pi/3$. (Compare this with the example at the top of page 81.)

Exercise 4E continued

3. On a single diagram draw and identify the locus of points z for which

$$\arg\left(\frac{z-3-2j}{z-1}\right) = \alpha \text{ where } x \text{ is}$$

(i) $\frac{\pi}{4}$ (ii) $-\frac{\pi}{4}$ (iii) $\frac{3\pi}{4}$ (iv) $-\frac{3\pi}{4}$

4. Prove that: $\arg\left(\frac{z-5}{z+5j}\right) = \pi/4 \Rightarrow |z| = 5$.

Investigate whether the converse is true.

5. Let A, B, C, D, E, F be the points representing a, b, c, d, e, f in the Argand diagram.

Prove that triangles ABC, DEF are directly similar if and only if

$$ae + bf + cd = af + bd + ce.$$

Find in a similar form the condition for these triangles to have opposite similarity.

6. The points A, B, C in the Argand diagram represent the complex numbers a, b, c, and $a = (1 - \lambda)b + \lambda c$. Prove that if λ is real then A lies on BC and divides BC in the ratio $\lambda : 1-\lambda$, but if λ is complex then, in triangle ABC, AB : BC = $|\lambda| : 1$ and angle ABC = $\arg \lambda$.

7. (i) If $\omega = \cos(2\pi/3) + j\sin(2\pi/3)$ and z is any vector, how are the vectors z and ωz related geometrically?

(ii) If $2 + 3j$ and $4 + 7j$ are two vertices of an equilateral triangle, find both possible positions for the third vertex.

8. (i) If the points a and b are two vertices of an equilateral triangle, prove that the third vertex is either $b + \omega(b - a)$ or $b + \omega^2(b - a)$, where ω is as in question 7.

(ii) Show that these expressions can be written as $-\omega a - \omega^2 b$ and $-\omega^2 a - \omega b$ respectively.

(iii) Deduce that the triangle with vertices z_1, z_2, z_3 is equilateral if and only if

$$z_1 + \omega z_2 + \omega^2 z_3 = 0 \text{ or } z_1 + \omega^2 z_2 + \omega z_3 = 0.$$

(iv) Deduce that a necessary and sufficient condition for the points z_1, z_2, z_3 to form an equilateral triangle is

$$z_1^2 + z_2^2 + z_3^2 = z_2 z_3 + z_3 z_1 + z_1 z_2.$$

9. The points A, B, C in the Argand diagram represent the complex numbers a, b, c; M is the midpoint of AB, and G is the point dividing the median AM in the ratio 2:1. Show that G represents the number $(a + b + c)/3$, and deduce from the symmetry of this expression that G also lies on the median through B and the median through C. (A *median* of a triangle is a line joining a vertex to the midpoint of the opposite side; the point G at which the medians meet is called the *centroid* of the triangle).

10. Directly similar triangles BCL, CAM, ABN are drawn on the sides of a triangle ABC. Prove that triangles ABC, LMN have the same centroid.

11. (i) On the sides of any triangle, equilateral triangles are drawn, pointing outward. Using question 8 part (ii), prove that the centroids of these equilateral triangles form another equilateral triangle. This is *Napoleon's Theorem*; it was attributed to the Emperor within a few years of his death, and he was a good enough mathematician to have discovered it.

(ii) Prove that the theorem is still true if the equilateral triangles are drawn inward rather than outward.

(iii) Prove that the triangle of centroids in (i), the corresponding triangle in (ii), and the original triangle all have the same centroid.

12. (i) Squares whose centres are P, Q, R, S are drawn outwards on the sides AB, BC, CD, DA of a general quadrilateral ABCD. Prove that PR and QS are equal and mutually perpendicular.

(ii) What difference does it make if all the squares are drawn inwards?

(iii) Explain how the result of the example on page 82 can be deduced from (i).

13. Directly similar triangles ABP, BCQ, CDR, DAS are drawn outwards on the sides of a general quadrilateral ABCD. Investigate the circumstances in which PR and QS are equal and mutually perpendicular.

Complex exponents

When multiplying complex numbers in polar form you add the arguments, and when multiplying powers of the same base you add the exponents. This suggests that there may be a link between the familiar expression $\cos\theta + j\sin\theta$ and the seemingly remote territory of the exponential function. This was first noticed in 1714 by the young Englishman Roger Cotes, two years before his death at the age of 28 (when Newton remarked 'If Cotes had lived we might have known something'), and made widely known through an influential book published by Euler in 1748.

If θ is any real number then

$$\cos\theta = 1 - \frac{\theta^2}{2!} + \frac{\theta^4}{4!} - \frac{\theta^6}{6!} + \ldots$$

and $$\sin\theta = \theta - \frac{\theta^3}{3!} + \frac{\theta^5}{5!} - \frac{\theta^7}{7!} + \ldots ,$$

and so $$\cos\theta + j\sin\theta = 1 - \frac{\theta^2}{2!} + \frac{\theta^4}{4!} - \frac{\theta^6}{6!} + \ldots + j\left(\theta - \frac{\theta^3}{3!} + \frac{\theta^5}{5!} - \frac{\theta^7}{7!} + \ldots\right).$$

Rearranging these two infinite series in ascending powers of θ gives

$$\begin{aligned}\cos\theta + j\sin\theta &= 1 + j\theta - \frac{\theta^2}{2!} - \frac{j\theta^3}{3!} + \frac{\theta^4}{4!} + \frac{j\theta^5}{5!} - \frac{\theta^6}{6!} - \frac{j\theta^7}{7!} + \ldots \\ &= 1 + j\theta + \frac{(j\theta)^2}{2!} + \frac{(j\theta)^3}{3!} + \frac{(j\theta)^4}{4!} + \frac{(j\theta)^5}{5!} + \frac{(j\theta)^6}{6!} + \frac{(j\theta)^7}{7!} + \ldots,\end{aligned}$$

which is what we get if we use $j\theta$ in place of x in the exponential series

$$e^x = 1 + x + \frac{x^2}{2!} + \frac{x^3}{3!} + \frac{x^4}{4!} + \ldots .$$

The problem here is that you do not know whether rearranging the terms of an infinite series can affect its sum. The answer is 'Yes sometimes, but not with the particular series involved here', but to prove this is beyond the scope of this course. This actually does not matter, because no meaning has yet been given to e^z when z is complex, and so we can make the following *definition*, suggested by this work with series but not dependent on it:

$$e^{j\theta} = \cos\theta + j\sin\theta.$$

NOTE

The particular case when $\theta = \pi$ gives $e^{j\pi} = \cos\pi + j\sin\pi = -1$, so that

$$e^{j\pi} + 1 = 0.$$

This remarkable statement linking the five fundamental numbers 0, 1, j, e, π, the three fundamental operations of addition, multiplication and exponentiation, and the fundamental relation of equality has been described as a 'mathematical poem'.

The first use of $e^{j\theta}$ is simply as a more compact way of writing familiar expressions. For example, the polar form $r(\cos\theta + j\sin\theta)$ can now be abbreviated to $re^{j\theta}$, and de Moivre's Theorem becomes the seemingly obvious statement

$$(e^{j\theta})^n = e^{jn\theta} \text{ for all rational } n.$$

The definition of e^z for any complex number z is now fairly obvious. Since we naturally want to preserve the basic property $e^{a+b} = e^a \times e^b$, it follows that if $z = x + jy$ then $e^z = e^x \times e^{jy}$.

Therefore we make the *definition*

$$e^z = e^x(\cos y + j\sin y).$$

Notice that when $y = 0$, $e^z = e^x$, so that when z is real this definition of e^z gives the exponential function we have used until now. Also, taking $x = 0$, $e^{jy} = \cos y + j\sin y$, agreeing with the definition suggested by the power series.

Activity

Prove that $e^{z+2n\pi j} = e^z$. This means that the exponential function is periodic, with the imaginary period $2\pi j$.

EXAMPLE

Given two complex numbers, z and w, prove from the definition that

$$e^{z+w} = e^z \times e^w.$$

Solution

Let $z = x + jy$ and $w = u + jv$. Then

$$\begin{aligned} e^z \times e^w &= e^x(\cos y + j\sin y) \times e^u(\cos v + j\sin v) \\ &= e^x e^u(\cos y + j\sin y)(\cos v + j\sin v) \\ &= e^{x+u}(\cos(y + v) + j\sin(y + v)) \\ &= e^{x+u+j(y+v)} \\ &= e^{z+w}. \end{aligned}$$

Since $e^{j\theta} = \cos\theta + j\sin\theta$ and $e^{-j\theta} = \cos(-\theta) + j\sin(-\theta) = \cos\theta - j\sin\theta$, it follows that

$$\cos\theta = \frac{e^{j\theta} + e^{-j\theta}}{2} \quad \text{and} \quad \sin\theta = \frac{e^{j\theta} - e^{-j\theta}}{2j}.$$

These are essentially the same as the results we made use of in the example on page 73.

Finally we can use these results for real θ to suggest natural definitions of the trigonometric functions for complex numbers z:

$$\cos z = \frac{e^{jz} + e^{-jz}}{2} \quad \text{and} \quad \sin z = \frac{e^{jz} - e^{-jz}}{2j}.$$

There is more to be said about this topic in Chapter 6 (page 131).

Exercise 4F

1 – 4. Express e^z in the form $x + jy$, where z is the given complex number.

1. $-j\pi$

2. $j\pi/4$

3. $(2 + 5j\pi)/6$

4. $3 - 4j$

5. Find all the solutions of $e^z = e^3$, and plot some of them on the Argand diagram.

6. Find all the solutions of $e^z = (1 - \sqrt{3}j)/2e^4$, and plot some of them on the Argand diagram.

7. Find all the values of z for which $e^{z^*} = (e^z)^*$.

8. Prove that $(1 + e^{2j\theta})^n = 2^n \cos^n\theta e^{jn\theta}$.

9. (i) Write down, in the form $a + jb$, the following complex numbers: $e^{j\theta}$, $e^{jn\theta}$ and $e^{-jn\theta}$.

(ii) Show that $(1 - \frac{1}{2}e^{2j\theta})(1 - \frac{1}{2}e^{-2j\theta}) = \frac{5}{4} - \cos 2\theta$.

The infinite series C and S are defined as follows:

$$C = \cos\theta + \tfrac{1}{2}\cos 3\theta + \tfrac{1}{4}\cos 5\theta + \tfrac{1}{8}\cos 7\theta + \ldots + \frac{1}{2^{r-1}}\cos(2r-1)\theta + \ldots$$

$$S = \sin\theta + \tfrac{1}{2}\sin 3\theta + \tfrac{1}{4}\sin 5\theta + \tfrac{1}{8}\sin 7\theta + \ldots + \frac{1}{2^{r-1}}\sin(2r-1)\theta + \ldots .$$

(iii) Show that $C + jS = \dfrac{4e^{j\theta} - 2e^{-j\theta}}{5 - 4\cos 2\theta}$.

(iv) Hence find expressions for C and S in terms of $\cos\theta$, $\sin\theta$ and $\cos 2\theta$ only.

[MEI]

10. If $z = f(p) + jg(p)$, where p is a real parameter then the derivative and integral of z with respect to p are defined by $\dfrac{dz}{dp} = f'(p) + jg'(p)$

and $\displaystyle\int z\,dp = \int f(p)\,dp + j\int g(p)\,dp$.

Prove that if $z = e^{\alpha p}$ where α is a fixed complex number, then $\dfrac{dz}{dp} = \alpha e^{\alpha p}$ and $\displaystyle\int z\,dp = \frac{e^{\alpha p}}{\alpha} + c$.

11. The position at time t of a point Z moving in the Argand diagram is given by $z = re^{j\theta}$, where r and θ depend on t. Find $\dfrac{dz}{dt}$ and $\dfrac{d^2z}{dt^2}$, and deduce the radial and transverse components of the velocity and acceleration of Z. (The *radial* and *transverse* directions are respectively parallel and perpendicular to $\overrightarrow{OZ}$.)

12. Let $C = \displaystyle\int e^{3x}\cos 2x\,dx$ and $S = \displaystyle\int e^{3x}\sin 2x\,dx$.

Show that $C + jS = e^{(3+2j)x}/(3 + 2j) + A$, where A is a constant. Hence find C and S.

13. Find $\displaystyle\int e^{ax}\cos bx\,dx$ and $\displaystyle\int e^{ax}\sin bx\,dx$

(i) by using integration by parts twice;

(ii) by using the method of Question 12.

Which method do you prefer?

14. (i) Prove from the definitions given on page 86 that

(a) $\cos^2 z + \sin^2 z = 1$;

(b) $\sin(z + w) = \sin z \cos w + \cos z \sin w$ (start with the right hand side).

(ii) Prove similarly one other standard trigonometric formula of your choice.

15. Prove that $w = \ln|z| + j\arg z$ is one solution of the equation $e^w = z$, and find all the other solutions in terms of z. The given solution is called the *principal logarithm* of z, and is written $\ln z$.

16. Find the principal logarithm of
(i) -1; (ii) j; (iii) $-5j$; (iv) $4 + 3j$.

17. A complex power of a non-zero complex number is defined as follows:

$$z^w = e^{w\ln z} \quad (z \neq 0).$$

Prove that j^j is real, and find its value to three significant figures.

18. Working to three significant figures, find

(i) 1^j; (ii) $(-1)^j$;

(iii) $(1 - j)^{1+j}$; (iv) $(1 + 2j)^{3+4j}$.

19. Describe the motion of the point representing z^j in the Argand diagram as the point representing z moves once clockwise around the unit circle, starting at -1.

Investigation

Roberts' Theorem. Figure 4.4 shows four rods AB, BC, CD, DA which are flexibly linked. Rod AD is fixed (sometimes the points A and D are just fixed without being joined by a rod), and a triangle BCP is attached to rod BC.

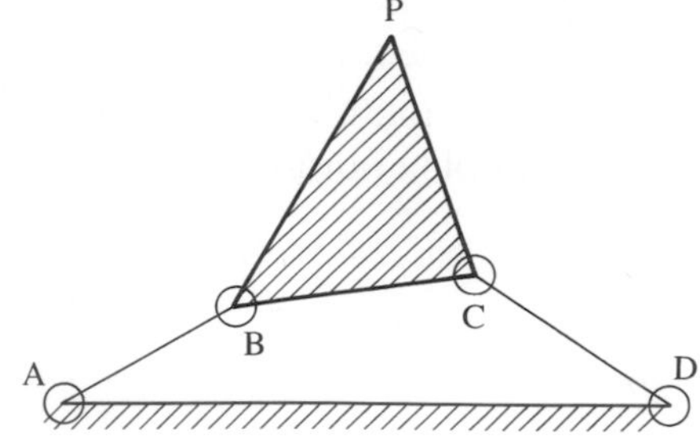

Figure 4.4

This mechanism is called a *four bar linkage*: by adjusting the lengths of the rods and the shape of the triangle it is possible to achieve many different paths for the point P as the mechanism moves. Four bar linkages are used to control the motion of parts of many machines (a good collection of examples is given in *Mathematics Meets Technology* by Brian Bolt, CUP 1991). In 1878 the English engineer Richard Roberts proved that any motion of P which can be produced by a particular four bar linkage can also be produced by two other linkages; this is useful since the other linkages may be more convenient to fit into the machine.

To prove Roberts' Theorem we complete the parallelograms ABPE, DCPF, then construct triangles EPG, PFH directly similar to triangle BCP, and finally complete parallelogram GPHK [figure 4.5].

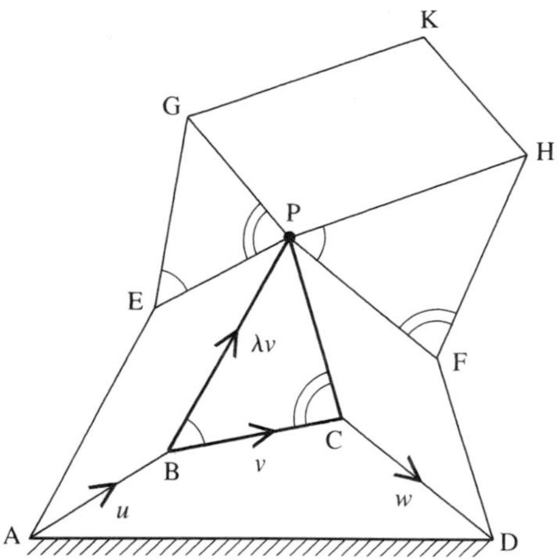

Figure 4.5

Putting the figure on the Argand diagram, let $\overrightarrow{AB}$, $\overrightarrow{BC}$, $\overrightarrow{CD}$ represent the complex numbers u, v, w respectively, and let the shape of triangle BCP be defined by the complex number λ, so that $\overrightarrow{BP} = \lambda v$.

Copy figure 4.5 and mark on each edge the complex number it represents.

Deduce that $\overrightarrow{AK}$ represents $\lambda(u + v + w)$, and hence that K is a fixed point. This shows that the linkage AEGK with triangle EGP and linkage DFHK with triangle FHP also give the same motion for P.

KEY POINTS

- de Moivre's Theorem: $(\cos\theta + j\sin\theta)^n = \cos n\theta + j\sin n\theta$, where n is rational.
- If $z = \cos\theta + j\sin\theta$ then $1/z = \cos\theta - j\sin\theta$,

  $\cos n\theta = (z^n + z^{-n})/2$, $\sin n\theta = (z^n - z^{-n})/2j$.
- The equation $z^n = 1$ has precisely n roots.

  These are $\omega^k = \cos\dfrac{2k\pi}{n} + j\sin\dfrac{2k\pi}{n}$, $k = 0, 1, 2, \ldots, n-1$.

  The sum of all these nth roots of unity is 0.
- The non-zero complex number $s(\cos\phi + j\sin\phi)$ has precisely n different nth roots. These are $s^{1/n}\left(\cos\left(\dfrac{\phi+2k\pi}{n}\right) + j\sin\left(\dfrac{\phi+2k\pi}{n}\right)\right)$, where $k = 0, 1, 2, \ldots, n-1$.

  The sum of these n roots is zero, and in the Argand diagram they are the vertices of a regular n-gon with centre O.
- In the Argand diagram if $\dfrac{a-b}{c-b} = \lambda$ then

  angle ABC $= \arg\lambda = \arg\left(\dfrac{a-b}{c-b}\right)$,

  and traingle ABC is similar to triangle ΛOI with vertices λ, 0, 1.
- $e^{j\theta} = \cos\theta + j\sin\theta$, $\cos n\theta = \dfrac{e^{jn\theta} + e^{-jn\theta}}{2}$, $\sin n\theta = \dfrac{e^{jn\theta} - e^{-jn\theta}}{2j}$.
- $e^z = e^x(\cos y + j\sin y)$, $\cos z = \dfrac{e^{jz} + e^{-jz}}{2}$, $\sin z = \dfrac{e^{jz} - e^{-jz}}{2j}$.

Conics

'Undoubtedly the record for the pay-back of a mathematical theory is held by the humble ellipse. Studied by the Greeks it really came into its own when planetary orbits were fully understood more than 1000 years later.'

Sir Michael Atiyah, President of the Royal Society, 1994

Each cable in the main span of a suspension bridge forms a parabola

The parabola

In Chapter 2 you proved some properties of conics in general from the polar equation $\frac{\ell}{r} = 1 + e \cos \theta$. Now we look in more detail at each type, starting with the parabola, for which $e = 1$.

The initial line is called the *axis* of the parabola, and the point O where the parabola meets its axis is called the *vertex*. Let the distance SO between the focus and the vertex be a; then putting $\theta = 0$ in $\frac{\ell}{r} = 1 + \cos \theta$ gives $\ell = 2a$ (figure 5.1).

The standard Cartesian equation uses the axis of the parabola as the x axis, with the origin at the vertex O. Until now we have placed the focus to the left of the vertex, but the parabola is traditionally drawn so that it faces the opposite way, open to the right. Then the focus S is $(a, 0)$ and the directrix d is $x = -a$ (figure 5.2).

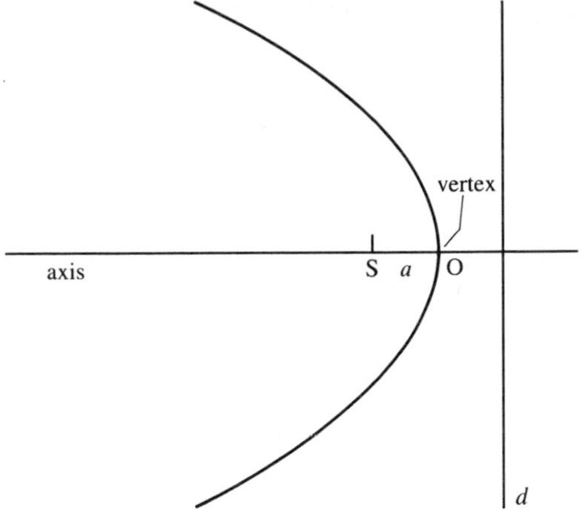

Figure 5.1

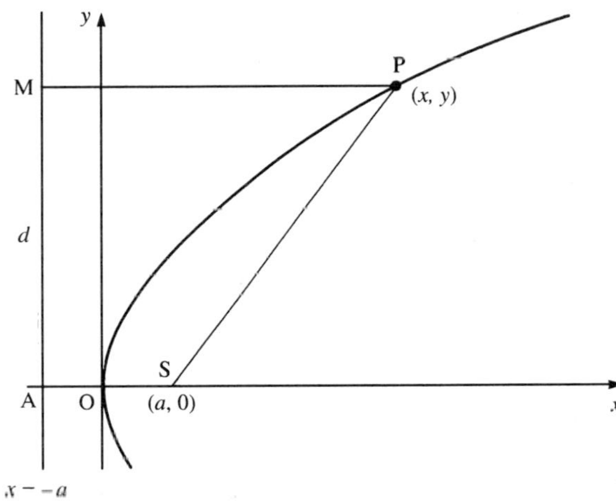

Figure 5.2

The point $P(x, y)$ is on the parabola $\Leftrightarrow$ $SP = PM$ (since $e = 1$)

$$\Leftrightarrow \quad SP^2 = PM^2$$

$$\Leftrightarrow \quad (x - a)^2 + y^2 = (x + a)^2$$

$$\Leftrightarrow \quad y^2 = (x + a)^2 - (x - a)^2$$

$$\Leftrightarrow \quad y^2 = 4ax.$$

This is the Cartesian equation of the parabola in its standard form.

The distance a determines the size of the parabola, but all parabolas are the same shape. In particular taking $a = \frac{1}{4}$ and then interchanging the x and y axes shows that the familiar curve with equation $y = x^2$ is a parabola with focus $(0, \frac{1}{4})$.

The parabola $y^2 = 4ax$ has simple parametric equations. Take (x, y) as any point on this parabola, and let $t = \frac{y}{2a}$; then $x = \frac{y^2}{4a} = \frac{4a^2t^2}{4a} = at^2$. So the parabola has parametric equations $x = at^2$, $y = 2at$, and all the points of the parabola can be given in the form $(at^2, 2at)$, with a unique value of t for each point.

Activity

Choose a convenient size for a and mark on a diagram the position of $(at^2, 2at)$ for $t = -3, -2, -1, -\frac{1}{2}, 0, \frac{1}{2}, 1, 2, 3$. Notice how the point moves along the parabola as t increases. If t changes at a constant rate does the point move with constant speed?

EXAMPLE

The points T and U of the parabola $y^2 = 4ax$ have co-ordinates $(at^2, 2at)$ and $(au^2, 2au)$ respectively. Find the equation of

(i) the chord TU;

(ii) the tangent to the parabola at T;

(iii) the normal to the parabola at T.

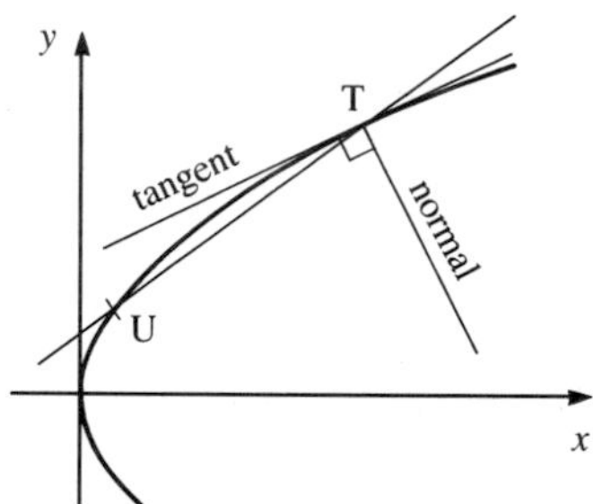

Solution

(i) The gradient of TU is $\frac{2at - 2au}{at^2 - au^2} = \frac{2(t-u)}{t^2 - u^2} = \frac{2}{t+u}$.
The line through T with this gradient has the equation

$$y - 2at = \frac{2}{t+u}(x - at^2)$$

$$\Leftrightarrow \quad (t + u)y - 2at\,(t + u) = 2x - 2at^2$$

$$\Leftrightarrow \quad 2x - (t + u)y + 2atu = 0.$$

(Note that this is symmetrical in t and u, as the geometry demands.)

(ii) The tangent at T is the limiting position of the chord TU as U $\to$ T along the curve, i.e. as $u \to t$. Letting $u \to t$ in the equation of the chord (and cancelling the common factor 2) gives the equation of the tangent:

$$x - ty + at^2 = 0.$$

(iii) From (ii) the gradient of the tangent at T is $1/t$, so the gradient of the normal is $-t$. The equation of the normal is

$$y - 2at = -t(x - at^2)$$

$$\Leftrightarrow \quad tx + y - 2at - at^3 = 0.$$

Activity

Prove these results about the gradient and equation of the tangent again using calculus.

Activity

Prove that TU is a focal chord if and only if $tu = -1$ (Reminder: a focal chord is a chord which passes through the focus.)

EXAMPLE

Prove that the tangents at the ends of a focal chord meet at right angles on the directrix.

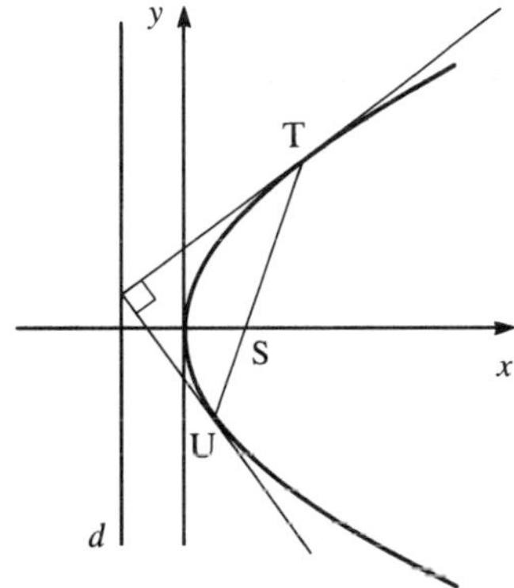

Solution

From the previous two activities the gradients of the tangents at T and U are $1/t$ and $1/u$, and $tu = -1$. Therefore the product of the gradients is -1, and so the tangents are perpendicular.

We already know from Question 10 of Exercise 2D that the tangents meet on the directrix, but we prove this again now to illustrate another method.

The equations of the tangents are

$$x - ty + at^2 = 0, \qquad ①$$

$$x - uy + au^2 = 0. \qquad ②$$

Taking ① $\times u$ – ② $\times t$ to eliminate y gives

$$(u - t)x + at^2u - au^2t = 0$$

$$\Leftrightarrow \quad (u - t)x = atu(u - t)$$

$$\Leftrightarrow \quad x = atu = -a \text{ since } tu = -1.$$

Therefore the tangents meet on the directrix.

Exercise 5A

Throughout this exercise $\boldsymbol{P}$ is the parabola with Cartesian equation $y^2 = 4ax$.

1. The diagram below shows a double square STUVU′T′ together with O, the midpoint of SV. Prove that the parabola with focus S and directrix UVU′ passes through T, O and T′, and touches VT and VT′. (Sketching or imagining this double square helps to give a reasonably correct shape when sketching a parabola.)

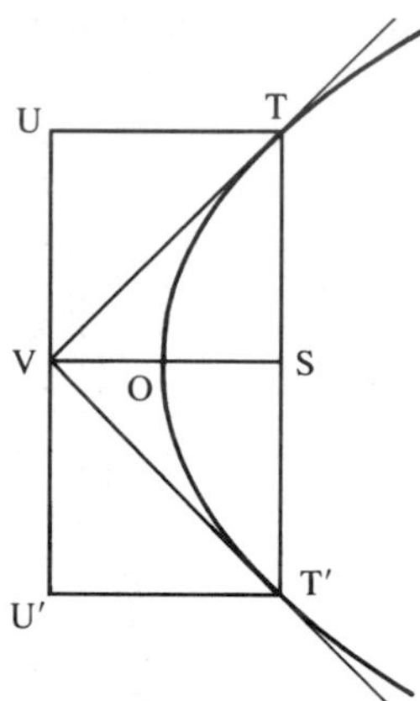

2. Draw a quick sketch of each of these parabolas, giving the co-ordinates of the vertex and focus, and the equations of the axis and directrix.

(i) $y^2 = 12x$

(ii) $y^2 = -x$

(iii) $(y - 2)^2 = 4(x - 3)$

(iv) $6y = x^2$

(v) $y^2 + 10y = 8x - 9$

(vi) $2y + x^2 + 6x - 1 = 0$

3. Find the equation of the parabola

(i) with focus $(5, 7)$ and directrix $x = -3$;

(ii) with vertex $(-4, 2)$ and directrix $x = 6$;

(iii) with vertex $(4, -1)$ and focus $(4, 7)$.

4. The point P of $\boldsymbol{P}$ has co-ordinates $(at^2, 2at)$; S is the focus, and M is the foot of the perpendicular from P to the directrix. The tangent at P meets SM at Z. Draw a diagram and prove that

(i) PZ is the perpendicular bisector of SM;

(ii) Z is on the tangent at O;

(iii) PZ bisects angle SPM.

5. Take a rectangular sheet of paper and mark a point S about 4 cm from the left-hand edge and about halfway down the sheet. Fold the paper so that this edge passes through S, make a sharp crease, and then unfold the paper. Do this repeatedly, always using the same edge but changing the position of the crease slightly each time. Use Question 4 part (i) to prove that all these creases touch a parabola, and identify its focus and directrix.

6. Draw a straight line m and a point S not on it. Place a set square with the right angle on m and one arm of the right angle through S. Draw the *other* arm of the right angle. Do this repeatedly, moving the right angle slightly along m each time but always keeping one arm through S. When necessary turn the set square over and move the right angle in the other direction along m. Use Question 4 part (ii) to prove that all these lines touch a parabola, and identify its focus and vertex.

7. Find the finite area of the region bounded by $\boldsymbol{P}$ and its latus rectum.

8. The tangent to $\boldsymbol{P}$ at P meets the directrix at R. Prove that PR subtends a right angle at the focus.

9. (i) Find the co-ordinates of the point where the tangents to $\boldsymbol{P}$ at $(ap^2, 2ap)$ and $(aq^2, 2aq)$ intersect.

(ii) If the two points on $\boldsymbol{P}$ in (i) vary in such a way that the direction of the chord joining them is fixed, prove that $p + q$ is constant.

(iii) Deduce that the midpoints of a set of parallel chords of $\boldsymbol{P}$ lie on a line parallel to the axis. Such a line is called a *diameter* of the parabola.

(iv) Prove that the tangents at the ends of any one of these parallel chords meet on this diameter.

10. P, Q, R, P′, Q′, R′ are points on $\boldsymbol{P}$ such that PQ′ is parallel to P′Q and QR′ is parallel to Q′R. Prove that RP′ is parallel to R′P. [**Hint:** write down conditions in terms of the parameters of these points for the given chords to be parallel.]

Exercise 5A continued

11. Prove that a circle which has a focal chord of a parabola as diameter touches the directrix.

12. The tangent and normal at a point P of $\mathcal{P}$ meet the axis at T and G respectively, and N is the foot of the perpendicular from P to the axis. Prove that TN is bisected by the vertex, and that NG is constant.

13. With the notation of Question 12, find the locus of

(i) the midpoint of OP as P varies;

(ii) the midpoint of PG as P varies;

(iii) the midpoint of a variable focal chord.

14. If the normal to $\mathcal{P}$ at $(at^2, 2at)$ meets $\mathcal{P}$ again at $(au^2, 2au)$, show that $t^2 + tu + 2 = 0$, and deduce that u^2 cannot be less than 8.

The line $3y = 2x + 4a$ meets $\mathcal{P}$ at the points P and Q. Prove that the normals at P and Q meet on $\mathcal{P}$.

15. The points P, Q, R of $\mathcal{P}$ have parameters p, q, r respectively. The tangents at Q and R meet at U, the tangents at R and P meet at V, and the tangents at P and Q meet at W.

(i) Find, in terms of p, q, r, the equation of the line through W perpendicular to UV.

(ii) Find the co-ordinates of the point where this line meets the directrix.

(iii) Deduce that the orthocentre of triangle UVW lies on the directrix.

(The *orthocentre* of a triangle is the point where the three perpendiculars from the vertices to the opposite edges meet.)

The ellipse

Next we investigate the ellipse, with polar equation $\frac{\ell}{r} = 1 + e\cos\theta$ where $0 < e < 1$.

The extreme values of r occur when $\theta = 0$ or π (figure 5.3):

$$\theta = 0 \Rightarrow r = \frac{\ell}{1+e} \text{ (at A say) and } \theta = \pi \Rightarrow r = \frac{\ell}{1-e} \text{ (at A}' \text{ say).}$$

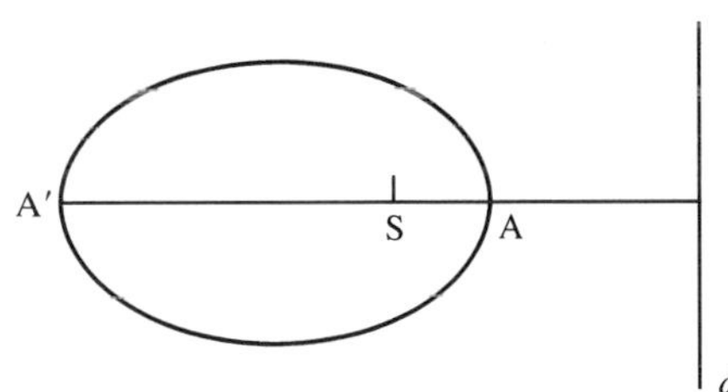

Figure 5.3

Now $AA' = \frac{\ell}{1+e} + \frac{\ell}{1-e} = \frac{2\ell}{1-e^2}$, so if $a = \frac{\ell}{1-e^2}$ then AA′ (which is called the *major axis*) is of length $2a$.

Next we find the Cartesian equation. Take the major axis as the x axis with the midpoint O of AA′ as the origin. Let OS = h and OX = k (figure 5.4).

Using the focus–directrix definition for the points A and A′ gives

$$SA = eAX, \quad \text{i.e.} \quad a - h = e(k - a),$$

$$SA' = eA'X, \quad \text{i.e.} \quad a + h = e(k + a),$$

from which $h = ae$ and $k = a/e$. Thus the focus S is $(ae, 0)$ and the directrix d is $x = a/e$.

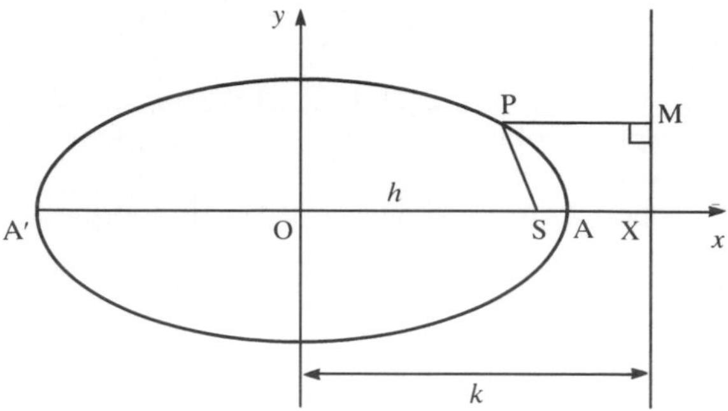

Figure 5.4

The point $P(x, y)$ is on the ellipse $\Leftrightarrow$ $SP = ePM$

$$\Leftrightarrow \quad SP^2 = e^2PM^2$$

$$\Leftrightarrow \quad (x - ae)^2 + y^2 = e^2\left(\frac{a}{e} - x\right)^2$$

$$\Leftrightarrow \quad x^2 - 2aex + a^2e^2 + y^2 = a^2 - 2aex + e^2x^2$$

$$\Leftrightarrow \quad x^2(1 - e^2) + y^2 = a^2(1 - e^2)$$

$$\Leftrightarrow \quad \frac{x^2}{a^2} + \frac{y^2}{a^2(1-e^2)} = 1$$

$$\Leftrightarrow \quad \frac{x^2}{a^2} + \frac{y^2}{b^2} = 1, \text{ where } b^2 = a^2(1 - e^2).$$

This is the standard Cartesian equation of the ellipse, which you have already met in *Pure Mathematics 1*; what we have just done shows that the definition of an ellipse given there and the focus–directrix definition give the same curve.

Activity

Show that this ellipse meets the y axis at the points $B(0, b)$ and $B'(0, -b)$. BB′ is called the *minor axis* of the ellipse.

Activity

Find b in terms of ℓ and e, and find a/b in terms of e.

The standard equation involves only even powers of x and y. So if the point (p, q) is on the ellipse then so are the points $(p, -q)$, $(-p, q)$ and $(-p, -q)$. Therefore the ellipse has line symmetry about *both* co-ordinate axes, and also half-turn symmetry about the origin. The origin is naturally called the *centre* of the ellipse; every chord through the centre is bisected there, so any such chord is called a *diameter*. By symmetry in the y axis there is a second focus S′ at $(-ae, 0)$, and corresponding to this a second directrix d' with equation $x = -a/e$ (figure 5.5).

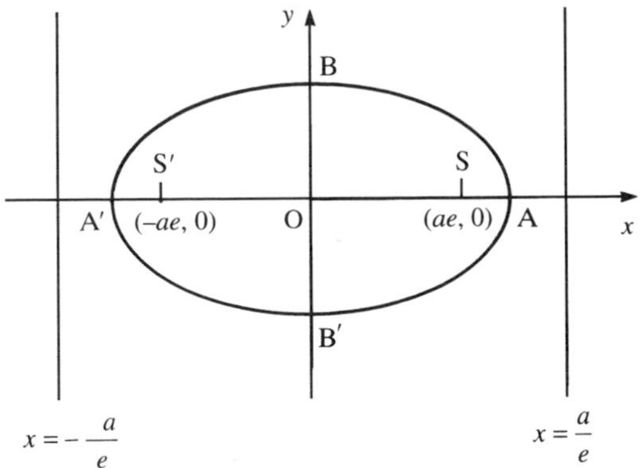

Figure 5.5

The number e fixes the position of each focus on the major axis. For example, if $e = \frac{2}{3}$ then each focus is two thirds of the way from the centre to an end of the major axis. This accounts for the name eccentricity (from the Latin for *from the centre*). If the eccentricity is very small then (from the preceding activity) a/b is close to 1, and the ellipse is nearly circular; indeed a circle can be thought of as an ellipse with zero eccentricity (where are the two foci and the two directrices?). If the eccentricity is close to 1 then a/b is large, and the ellipse is close to the straight line segment AA′.

Activity

Find the eccentricity of an ellipse with axes of lengths
(i) 10 cm, 6 cm; (ii) 50 cm, 48 cm; (iii) 1 m, 1 cm.

Activity

An ellipse has eccentricity $\frac{2}{3}$. Find, to the nearest centimetre,

(i) b and ℓ when $a = 100$ cm;
(ii) a and ℓ when $b = 100$ cm;
(iii) a and b when $\ell = 100$ cm.

Another way of obtaining an ellipse is by squashing a circle. To be more precise, consider the one-way stretch with matrix $\begin{pmatrix} 1 & 0 \\ 0 & b/a \end{pmatrix}$. This maps the point (X, Y) to the point (x, y) where $\begin{pmatrix} x \\ y \end{pmatrix} = \begin{pmatrix} 1 & 0 \\ 0 & b/a \end{pmatrix} \begin{pmatrix} X \\ Y \end{pmatrix} = \begin{pmatrix} X \\ bY/a \end{pmatrix}$, so that the x co-ordinate is unchanged, and the y co-ordinate is multiplied by b/a (which gives a 'squash' rather than a 'stretch' since $b/a < 1$):

$$x = X, \; y = \frac{bY}{a} \quad \text{or} \quad Y = \frac{ay}{b}.$$

Now apply this transformation to the circle with centre O and radius a (figure 5.6).

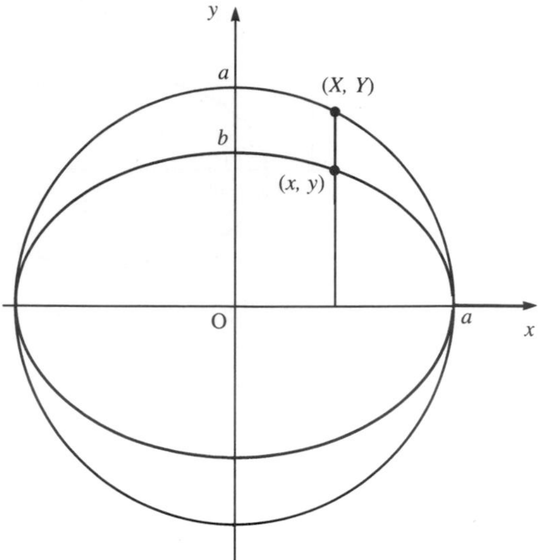

Figure 5.6

$$\begin{aligned}
(X, Y) \text{ is on this circle} \quad &\Leftrightarrow \quad X^2 + Y^2 = a^2 \\
&\Leftrightarrow \quad x^2 + \frac{a^2y^2}{b^2} = a^2 \\
&\Leftrightarrow \quad \frac{x^2}{a^2} + \frac{y^2}{b^2} = 1 \\
&\Leftrightarrow \quad (x, y) \text{ is on the ellipse.}
\end{aligned}$$

This shows that an ellipse is formed by shortening, in a given ratio, all the chords of a circle which are perpendicular to one diameter. When you look at a circular object you often see an ellipse because of your oblique viewpoint.

Activity

Find the determinant of $\begin{pmatrix} 1 & 0 \\ 0 & b/a \end{pmatrix}$. Deduce that the area of the ellipse is πab.

This 'squashing' transformation leads to the very useful standard parametric equations for the ellipse. If P_1 is the point on the circle with centre O and radius a such that angle $AOP_1 = \theta$ then P_1 has co-ordinates $(a\cos\theta, a\sin\theta)$ and the corresponding point P on the ellipse has co-ordinates $\left(a\cos\theta, \frac{b}{a} \times a\sin\theta\right)$,

i.e. $(a\cos\theta, b\sin\theta)$.

So the ellipse has parametric equations $x = a\cos\theta$, $y = b\sin\theta$ (figure 5.7).

As θ increases from 0 to 2π the point P moves once round the ellipse. The angle θ is called the *eccentric angle* of the point P; notice that this is *not* the angle AOP.

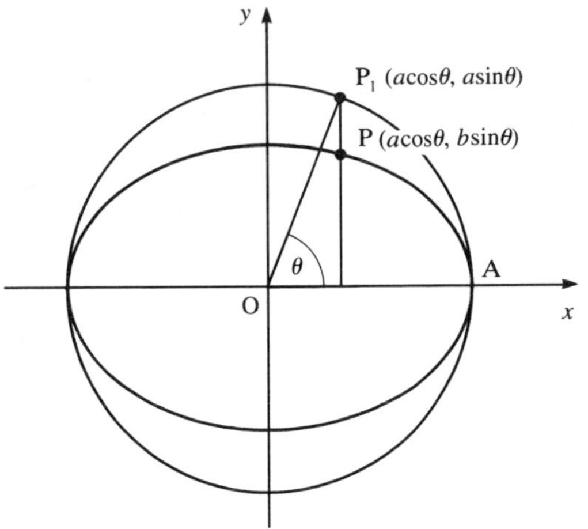

Figure 5.7

EXAMPLE

Find the equation of (i) the tangent, (ii) the normal to the standard ellipse at $P(a\cos\theta, b\sin\theta)$.

Solution

(i) $x = a\cos\theta \Rightarrow \dfrac{dx}{d\theta} = -a\sin\theta$ and $y = b\sin\theta \Rightarrow \dfrac{dy}{d\theta} = b\cos\theta$.

Therefore : $\dfrac{dy}{dx} = -\dfrac{b\cos\theta}{a\sin\theta}$ and the equation of the tangent at P is

$$y - b\sin\theta = -\frac{b\cos\theta}{a\sin\theta}(x - a\cos\theta)$$

$$\Leftrightarrow \quad ay\sin\theta - ab\sin^2\theta = -bx\cos\theta + ab\cos^2\theta$$

$$\Leftrightarrow \quad bx\cos\theta + ay\sin\theta = ab(\cos^2\theta + \sin^2\theta)$$

$$\Leftrightarrow \quad \frac{x\cos\theta}{a} + \frac{y\sin\theta}{b} = 1.$$

(ii) Using $mm' = -1$ for perpendicular lines

the gradient of the normal is $\dfrac{a\sin\theta}{b\cos\theta}$.

The equation of the normal at P is

$$y - b\sin\theta = \frac{a\sin\theta}{b\cos\theta}(x - a\cos\theta)$$

$$\Leftrightarrow \quad by\cos\theta - b^2\cos\theta\sin\theta = ax\sin\theta - a^2\cos\theta\sin\theta$$

$$\Leftrightarrow \quad ax\sin\theta - by\cos\theta = (a^2 - b^2)\cos\theta\sin\theta.$$

Exercise 5B

Throughout this exercise E is the ellipse with Cartesian equation $\frac{x^2}{a^2}+\frac{y^2}{b^2}=1$.

1. Draw the ellipse $\frac{x^2}{25}+\frac{y^2}{16}=1$ accurately on graph paper. Calculate the eccentricity, and add the foci and directrices to your diagram. Check the properties $SP = ePM$ and $S'P = ePM'$ by measurement from your diagram for three positions of the point P on this ellipse. (Note: *foci, directrices* are the plurals of *focus, directrix.*)

2. The orbit of the planet Pluto is an ellipse with major axis of length 1.18×10^{10} km and eccentricity $\frac{1}{4}$. Calculate the length of the minor axis.

3. Prove that the ratio of the least and greatest distances of a point of E from a focus is $1-e:1+e$.

4. The orbit of Halley's comet is an ellipse with the Sun at one focus. The distances of the perihelion and the aphelion from the Sun are 0.587 a.u. and 17.947 a.u. respectively. Find the eccentricity of the orbit. (The *perihelion* and *aphelion* are the points of an orbit closest to and furthest from the Sun; the *astronomical unit* (a.u.) is the mean distance of the Earth from the Sun, about 1.5×10^{11} m.)

5. The orbit of a satellite is an ellipse of eccentricity $\frac{1}{35}$ with the centre of the Earth at one focus. The Earth may be treated as a sphere of radius 6400 km. If the least height of the satellite above the Earth's surface is 400 km, what is the greatest height?

6. Explain why the distance d from the centre of the sun to the Earth is inversely proportional to the angle α which the Sun's diameter subtends at a point of the Earth (see below).

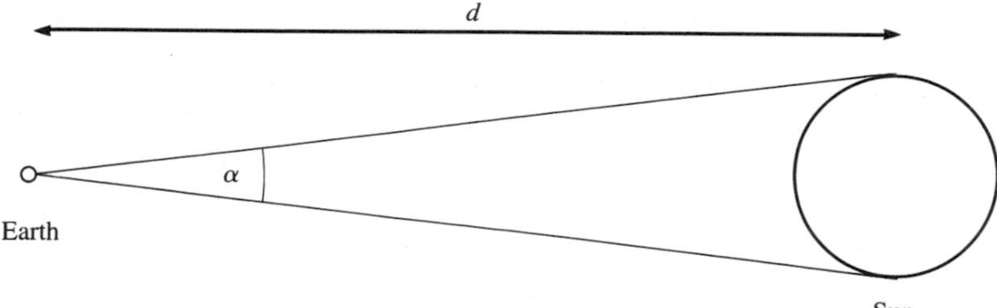

At the perihelion of the Earth's orbit this angle is 32′36″ and at the aphelion it is 31′32″. Calculate the eccentricity of the Earth's orbit. (The *minute* (′) and *second* (″) are used to measure small angles: $1° = 60'$, $1' = 60''$.)

7. Rearrange $2(x-1)^2 + 11(y-4)^2 = 22$ in the form $\frac{(x-h)^2}{a^2}+\frac{(y-k)^2}{b^2}=1$. Hence show that this is the equation of an ellipse. Find the coordinates of the centre and the foci, the equations of the directrices, and sketch the ellipse.

8. Repeat Question 7 for the equation $x^2 + 2y^2 + 4x - 8y + 4 = 0$.

9. The *elliptic trammel* is a mechanical device for drawing ellipses.

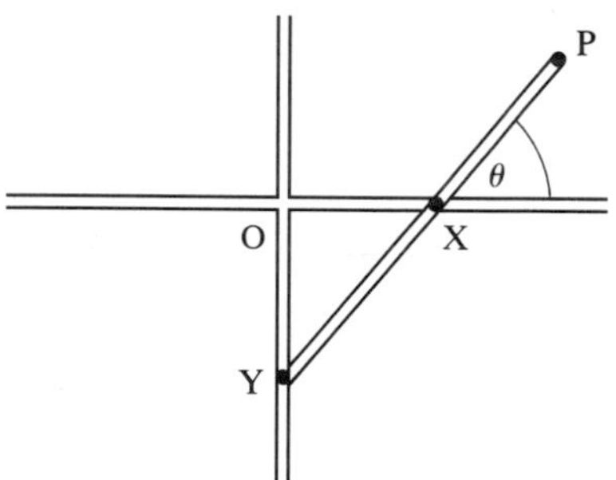

It consists of a straight rod with a pencil at P and pegs at X, Y which run in perpendicular grooves OX, OY. Prove that if OX, OY are taken as x and y axes with $PX = b$, $PY = a$, then the locus of P is E. (**Hint:** use the angle θ shown in the diagram.)

10. A ladder slides in a fixed vertical plane with its foot on the horizontal floor and its top against a vertical wall. A pot of paint is hooked onto the ladder, not at either end. Prove that the locus of the pot is part of an ellipse. Where is the pot if this ellipse is a circle?

Exercise 5B continued

11. The diagram below shows the plan of the door arrangement of a particular type of French telephone booth.

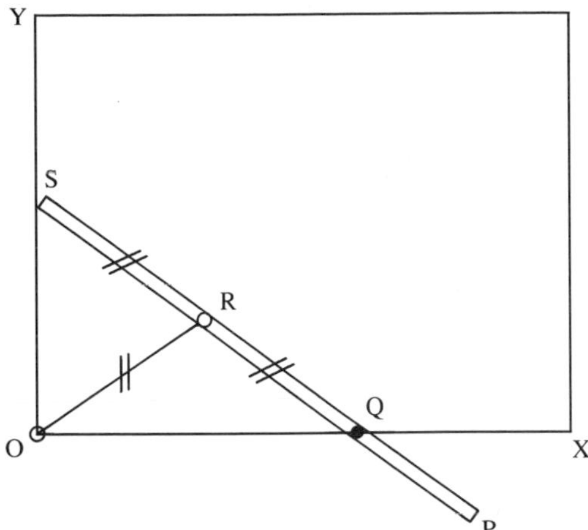

PQRS is the top of the door. The point Q moves in groove OX at the top of the door frame, and the point R is joined to the fixed point O by a horizontal link OR which is free to rotate at O and R (there are a similar groove and link at the bottom of the door). The lengths OR, SR, RQ are all equal. When the door is shut PQRS lies along OX, with S at O. When the door is fully open PQRS lies along OY with Q at O and P outside the booth.

(i) Prove that as the door moves S always lies on OY.

(ii) Prove that the path of P is quarter of an ellipse.

12. The line $y = mx + c$ meets the ellipse $\frac{x^2}{a^2}+\frac{y^2}{b^2}=1$ at two points P and Q.

(i) Show that the x co-ordinates of P and Q satisfy
$(a^2m^2 + b^2)x^2 + (2a^2mc)x + a^2(c^2 - b^2) = 0$.

(ii) Find the co-ordinates of the midpoint of PQ (in terms of a, b, m and c).

(iii) Show that the midpoint of PQ lies on the line $y=-\frac{b^2}{a^2m}x$.

Now suppose that P is the point $(a\cos\theta, b\sin\theta)$ and PQ is the normal to the ellipse at P.

(iv) Find the gradient of PQ in terms of a, b and θ.

(v) Given that the midpoint of PQ lies on the line $y = -x$, show that $\tan\theta=\frac{b^3}{a^3}$.

[MEI]

13. By using the condition for the roots of the quadratic equation of Question 12 part (i) to coincide, prove that the two tangents of $\boldsymbol{E}$ with gradient m have equations
$y = mx \pm \sqrt{a^2m^2+b^2}$.

14. Use Question 13 to show that the tangents from the point (–2, 5) to the ellipse $\frac{x^2}{6}+\frac{y^2}{3}=1$ have gradients –1 and 11. Find the equations of these two tangents. Find also the co-ordinates of the point of contact of each tangent.

15. (i) Use Question 13 to prove that the gradients of the two tangents from the point (X, Y) to $\boldsymbol{E}$ are the roots of the quadratic equation
$m^2(a^2 - X^2) + 2mXY + b^2 - Y^2 = 0$.

(ii) Find the condition for this equation to have complex roots, and interpret this geometrically.

(iii) Find the condition for the product of the roots to equal –1. Deduce that the tangents from the point (X, Y) to $\boldsymbol{E}$ are perpendicular if and only if (X, Y) lies on the circle $x^2 + y^2 = a^2 + b^2$. (This is called the *director circle* of the ellipse.)

16. An elliptical disc slides between two fixed perpendicular lines. Prove that the locus of its centre is an arc of a circle.

17. The point P of $\boldsymbol{E}$ has co-ordinates $(a\cos\theta, b\sin\theta)$. The normal at P meets the x axis at N, and the foot of the perpendicular from the origin O to the tangent at P is M. Prove that $OM \times PN = b^2$.

Exercise 5B continued

18. Conjugate diameters. If the eccentric angles of points P and U of $\boldsymbol{E}$ differ by $\pm\frac{\pi}{2}$ then the diameters PQ and UV are called *conjugate* diameters.

(i) Prove that when a circle is 'squashed' to form an ellipse [as on page 97], perpendicular diameters of the circle become conjugate diameters of the ellipse.

(ii) Prove that if conjugate diameters of $\boldsymbol{E}$ have gradients m, m' then $mm' = -b^2/a^2$.

(iii) Prove that, for conjugate diameters PQ and UV, $PQ^2 + UV^2 = 4(a^2 + b^2)$.

(iv) Prove that the locus of the midpoints of chords of $\boldsymbol{E}$ which are parallel to a diameter PQ is the conjugate diameter.

[**Hint:** 'unsquash' the ellipse back to a circle, noting that under this transformation midpoints of lines remain midpoints.]

(v) Prove that the area of the parallelogram formed by the four tangents at the ends of a pair of conjugate diameters is always $4ab$.

(vi) PQ is a diameter and R is any other point of an ellipse. Prove that the diameters parallel to PR and QR are conjugate.

(vii) A parallelogram is inscribed in an ellipse. Prove that its sides are parallel to a pair of conjugate diameters. Deduce that just one square can be inscribed in $\boldsymbol{E}$, and prove that its area is $\dfrac{4a^2b^2}{a^2+b^2}$.

The hyperbola

When $e > 1$ the conic with polar equation $\dfrac{\ell}{r} = 1 + e\cos\theta$ is a hyperbola, meeting the initial line where $\theta = 0$ or π:

$\theta = 0 \Rightarrow r = \dfrac{\ell}{1+e}$, at A say, and $\theta = \pi \Rightarrow r = \dfrac{\ell}{1-e}$, at A′.

Since $e > 1$, $\dfrac{\ell}{1-e}$ is negative, so A′ is to the right of S, with

$$|SA'| = -\frac{\ell}{1-e} = \frac{\ell}{e-1}.$$

Because $e - 1 < e + 1$ we also have $\dfrac{\ell}{e-1} > \dfrac{\ell}{e+1}$, so A′ is also to the right of A (figure 5.8) and $AA' = \dfrac{\ell}{e-1} - \dfrac{\ell}{e+1} = \dfrac{2\ell}{e^2-1}$.

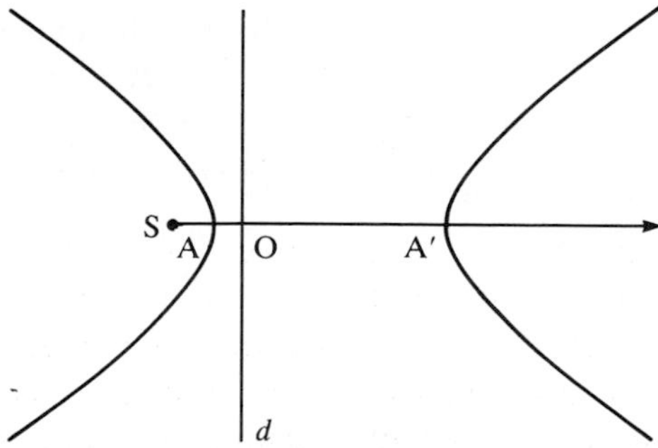

Figure 5.8

To find the Cartesian equation we take the origin O at the midpoint of AA′ with the initial line as the x axis. It is customary to reverse the diagram, so that S and A are on the positive x axis, as in figure 5.9. Then A has co-ordinates $(a, 0)$, where

$$a = \frac{\ell}{e^2 - 1} = \tfrac{1}{2}\text{AA}'.$$

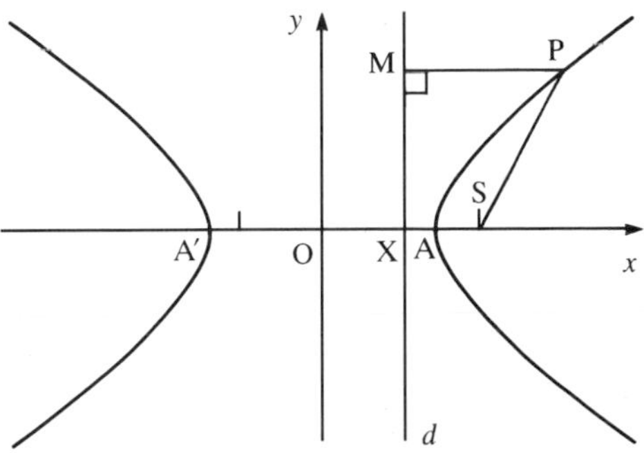

Figure 5.9

The pattern of argument is now the same as for the ellipse, so compare what follows with pages 95 and 96, noting the similarities and differences.

Let OS = h and OX = k. Then using the focus–directrix definition for the points A and A′ gives

$$\text{SA} = e\text{AX}, \qquad \text{i.e. } h - a = e(a - k),$$

$$\text{SA}' = e\text{A}'\text{X}, \qquad \text{i.e. } h + a = e(a + k),$$

from which $h = ae$ and $k = a/e$. Thus the focus S is $(ae, 0)$ and the directrix d is $x = a/e$.

The point P(x, y) is on the hyperbola $\Leftrightarrow$ SP = ePM

$\Leftrightarrow \quad \text{SP}^2 = e^2\text{PM}^2$

$\Leftrightarrow \quad (x - ae)^2 + y^2 = e^2\left(x - \dfrac{a}{e}\right)^2$

$\Leftrightarrow \quad x^2 - 2aex + a^2e^2 + y^2 = e^2x^2 - 2aex + a^2$

$\Leftrightarrow \quad x^2(1 - e^2) + y^2 = a^2(1 - e^2)$

$\Leftrightarrow \quad \dfrac{x^2}{a^2} + \dfrac{y^2}{a^2(1-e^2)} = 1$

It is convenient to change signs here since $a^2(e^2 - 1)$ is positive.

$\Leftrightarrow \quad \dfrac{x^2}{a^2} - \dfrac{y^2}{a^2(e^2-1)} = 1$

$\Leftrightarrow \quad \dfrac{x^2}{a^2} - \dfrac{y^2}{b^2} = 1$, where $b^2 = a^2(e^2 - 1)$.

This is the standard Cartesian equation of the hyperbola. The x axis is called the *transverse axis*, and the y axis (which the curve does not meet) is called the *conjugate axis*. Since the equation contains only even powers of x and y the hyperbola is symmetrical about both axes, so there are two foci, at $(\pm ae, 0)$, and two directrices with equations $x = \pm a/e$ (figure 5.10).

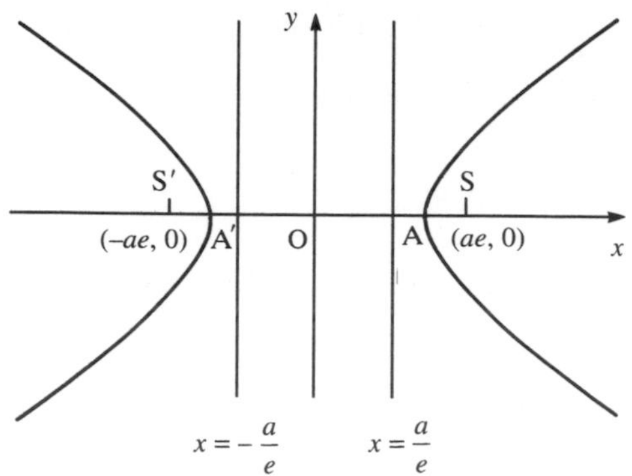

Figure 5.10

For many purposes the most convenient parametric equations for the hyperbola are

$$x = a\sec\theta\,,\; y = b\tan\theta.$$

Activity

Check that $(a\sec\theta,\, b\tan\theta)$ satisfies $\dfrac{x^2}{a^2} - \dfrac{y^2}{b^2} = 1$. Describe how this point moves once along the whole hyperbola as θ increases from 0 to 2π, paying particular attention to what happens when θ is near $\pi/2$ or $3\pi/2$.

Activity

Prove that, for points $(a\sec\theta,\, b\tan\theta)$ of the hyperbola, $\dfrac{y}{x} \to \dfrac{b}{a}$ as $\theta \to \dfrac{\pi}{2}$ and $\dfrac{y}{x} \to -\dfrac{b}{a}$ as $\theta \to \dfrac{3\pi}{2}$.

EXAMPLE

Prove that the equation of the tangent to the hyperbola at $P(a\sec\theta,\, b\tan\theta)$ may be written as $\dfrac{x\sec\theta}{a} - \dfrac{y\tan\theta}{b} = 1$ or as $\dfrac{x}{a} - \dfrac{y}{b}\sin\theta = \cos\theta$.

Solution

$x = a\sec\theta \Rightarrow \dfrac{dx}{d\theta} = a\sec\theta\tan\theta$ and $y = b\tan\theta \Rightarrow \dfrac{dy}{d\theta} = b\sec^2\theta$.

Therefore $\dfrac{dy}{dx} = \dfrac{b\sec^2\theta}{a\sec\theta\tan\theta} = \dfrac{b\sec\theta}{a\tan\theta}$ and the equation of the tangent at P is

$$y - b\tan\theta = \frac{b\sec\theta}{a\tan\theta}(x - a\sec\theta)$$

$$\Leftrightarrow \quad ay\tan\theta - ab\tan^2\theta = bx\sec\theta - ab\sec^2\theta$$

$$\Leftrightarrow \quad bx\sec\theta - ay\tan\theta = ab(\sec^2\theta - \tan^2\theta)$$

$$\Leftrightarrow \quad \frac{x\sec\theta}{a} - \frac{y\tan\theta}{b} = 1,\ \text{since } \sec^2\theta - \tan^2\theta = 1.$$

Multiplying throughout by $\cos\theta$ and using $\tan\theta = \dfrac{\sin\theta}{\cos\theta}$ gives the alternative form $\dfrac{x}{a} - \dfrac{y}{b}\sin\theta = \cos\theta$.

You have already seen in the first activity of this section that the point $P(a\sec\theta, b\tan\theta)$ moves to infinity as $\theta \to \dfrac{\pi}{2}$. The second form of the equation of the tangent given in the example above shows that as this happens the tangent at P approaches the *asymptote* $\dfrac{x}{a} - \dfrac{y}{b} = 0$. Similarly the other asymptote $\dfrac{x}{a} + \dfrac{y}{b} = 0$ is the limiting position of the tangent as $\theta \to \dfrac{3\pi}{2}$.

Activity

Show that the family of hyperbolas $\dfrac{x^2}{a^2} - \dfrac{y^2}{b^2} = k$, where a and b are fixed but k varies, all have the same asymptotes. Sketch the curves for the values $k = 2, 1, \frac{1}{2}, 0, -\frac{1}{2}, -1, -2$.

Activity

Prove that the acute angle between an asymptote and the transverse axis is arcsec e.

Because their equations are so similar, the ellipse and the hyperbola share many properties. The obvious main differences between them are that the hyperbola has two branches and two asymptotes, so the distinctive features of the hyperbola usually involve the asymptotes.

EXAMPLE

Lines are drawn parallel to the asymptotes through any point P of the hyperbola $\dfrac{x^2}{a^2} - \dfrac{y^2}{b^2} = 1$, meeting the asymptotes at H and K. Prove that $PH \times PK = \frac{1}{4}(a^2 + b^2)$. (This property will be useful in the next section.)

Solution

By symmetry there is no loss of generality in taking P in the first quadrant as in the diagram overleaf. Let $PH = h$, $PK = k$, and let the angle between the asymptote and the x axis be ϕ.

Then $\quad a\sec\theta = k\cos\phi + h\cos\phi \Rightarrow k + h = a\sec\theta\sec\phi$

and $\quad b\tan\theta = k\sin\phi - h\sin\phi \Rightarrow k - h = b\tan\theta\,\mathrm{cosec}\,\phi$

So $\quad 4hk = (k+h)^2 - (k-h)^2$

$$= a^2\sec^2\theta\sec^2\phi - b^2\tan^2\theta\,\mathrm{cosec}^2\phi$$

$$= (a\sec\phi)^2\sec^2\theta - (b\,\mathrm{cosec}\,\phi)^2\tan^2\theta.$$

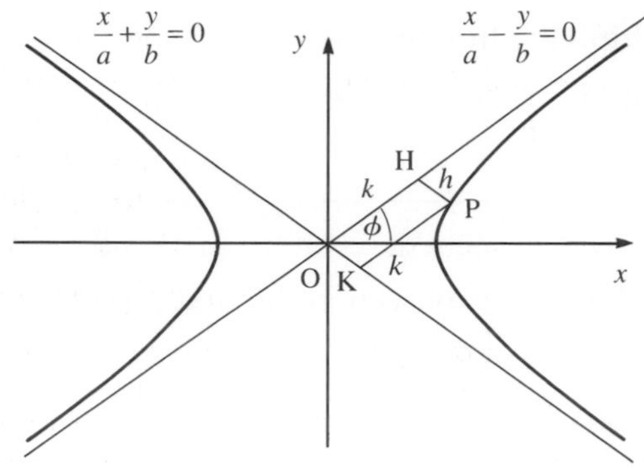

But since $\tan\phi = \dfrac{b}{a}$, $a\sec\phi = b\operatorname{cosec}\phi = \sqrt{a^2+b^2}$ (see the diagram below).

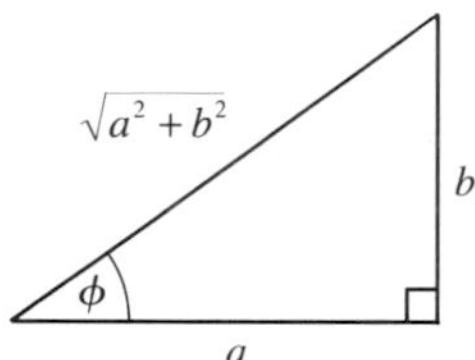

Therefore $\quad 4hk = (a^2 + b^2)(\sec^2\theta - \tan^2\theta) = (a^2 + b^2)$

and so $\quad \text{PH} \times \text{PK} = \frac{1}{4}(a^2 + b^2)$.

Exercise 5C

Throughout this exercise ***H*** is the hyperbola

$$\frac{x^2}{a^2} - \frac{y^2}{b^2} = 1.$$

1. The rectangle TUVW has vertices (a, b), $(-a, b)$, $(-a, -b)$, $(a, -b)$ respectively. Show that

(i) ***H*** touches TW and UV at their midpoints A and A′;

(ii) TV and UW are the asymptotes of ***H***;

(iii) the circle with TV as diameter meets the x axis at the foci of ***H***;

(iv) the directrices pass through the points where the circle with diameter AA′ meets the asymptotes.

2. Starting with the rectangle TUVW of Question 1 and using its properties draw quick sketches of the following hyperbolas with their asymptotes, foci and directrices.

(i) $\dfrac{x^2}{16} - \dfrac{y^2}{9} = 1$ (ii) $\dfrac{x^2}{9} - \dfrac{y^2}{16} = 1$

(iii) $\dfrac{x^2}{4} - y^2 = 1$ (iv) $x^2 - y^2 = 36$.

3. Find the equation of the hyperbola with foci $(\pm 8, 0)$ and directrices $x = \pm 2$.

4. Find the equation of the hyperbola which has asymptotes $y = \pm 3x$ and passes through $(2, 4)$. Find also the equations of the tangent and normal at $(2, 4)$.

5. The line $y = mx + c$ meets ***H*** at P_1, P_2 and meets the asymptotes at Q_1, Q_2.

(i) Write down the quadratic equation whose roots are the x co-ordinates of P_1, P_2, and find the sum of these roots.

(ii) Write down the quadratic equation whose roots are the x co-ordinates of Q_1, Q_2, and find the sum of these roots.

(iii) Hence show that P_1P_2 and Q_1Q_2 have the same midpoint.

(iv) Deduce that $P_1Q_1 = P_2Q_2$.

6. The tangent at a point P of $\boldsymbol{H}$ meets the asymptotes at Q_1, Q_2.

(i) Prove that P is the midpoint of Q_1Q_2. [**Hint:** use Question 5 part (iv).]

(ii) Prove that as P varies the area of triangle OQ_1Q_2 remains constant.

7. The hyperbolas $\boldsymbol{H}$ and $\frac{x^2}{a^2} - \frac{y^2}{b^2} = -1$ are said to be *conjugate*.

Sketch both these hyperbolas on a single diagram.

If their eccentricities are e and f, show that $e^2f^2 = e^2 + f^2$.

8. Prove that:
the line $y = mx + c$ touches $\boldsymbol{H} \Rightarrow a^2m^2 = b^2 + c^2$.

Investigate whether the converse is true.

9. Prove that the equation of the normal to $\boldsymbol{H}$ at $P(a\sec\theta, b\tan\theta)$ is $ax\sin\theta + by = (a^2 + b^2)\tan\theta$.

This normal meets the transverse axis at G, and the midpoint of PG is Q. Prove that the locus of Q is a hyperbola.

10. Adapt the method of Exercise 5B, Question 15, to show that the locus of the point of intersection of perpendicular tangents to $\boldsymbol{H}$ is the circle $x^2 + y^2 = a^2 - b^2$ if $a > b$, and that there are no perpendicular tangents if $a < b$. What happens when $a = b$?

11. (i) Prove that the lines $\frac{x}{a} + \frac{y}{b} = t$, $\frac{x}{a} - \frac{y}{b} = \frac{1}{t}$ $(t \neq 0)$, which are parallel to the asymptotes, meet on $\boldsymbol{H}$.

(ii) Deduce the alternative parametric equations $x = \frac{a}{2}\left(t + \frac{1}{t}\right), y = \frac{b}{2}\left(t - \frac{1}{t}\right)$ for $\boldsymbol{H}$.

(iii) The same point of $\boldsymbol{H}$ has co-ordinates $(a\sec\theta, b\tan\theta)$ and $\left(\frac{a}{2}\left(t + \frac{1}{t}\right), \frac{b}{2}\left(t - \frac{1}{t}\right)\right)$.

Prove that $t = \tan\left(\frac{\theta}{2} + \frac{\pi}{4}\right)$.

(iv) Show that the equation of the tangent to $\boldsymbol{H}$ at $\left(\frac{a}{2}\left(t + \frac{1}{t}\right), \frac{b}{2}\left(t - \frac{1}{t}\right)\right)$ is

$$\frac{x}{a}\left(t^2 + 1\right) - \frac{y}{b}\left(t^2 - 1\right) = 2t.$$

(v) Find where the tangent in (iv) meets the asymptotes, and deduce the results of Question 6 again.

The rectangular hyperbola

In the special case when $a = b$ the equation of the hyperbola may be written as $x^2 - y^2 = a^2$, and the asymptotes are the lines $y = \pm x$. Since these are at right angles the curve is called a *rectangular hyperbola* (figure 5.11).

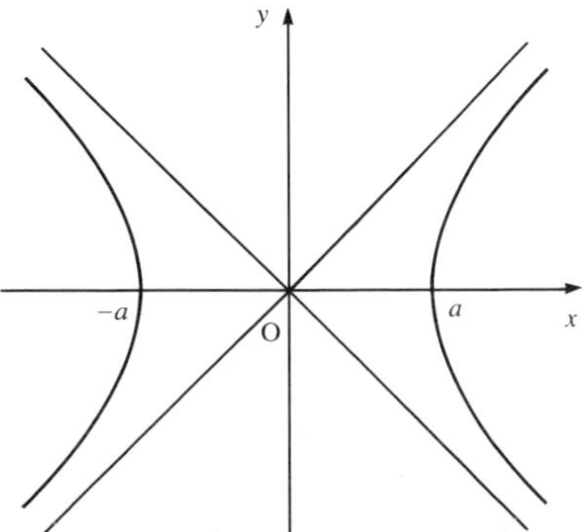

Figure 5.11

Activity

Show that all rectangular hyperbolas have eccentricity $\sqrt{2}$, and hence that they are all the same shape. (The relationship between rectangular hyperbolas and general hyperbolas is like that between circles and ellipses.)

Since the asymptotes of a rectangular hyperbola are perpendicular they can be used as co-ordinate axes. Let $P(x, y)$ be a point on a rectangular hyperbola referred to these axes, with one branch in the first quadrant (figure 5.12).

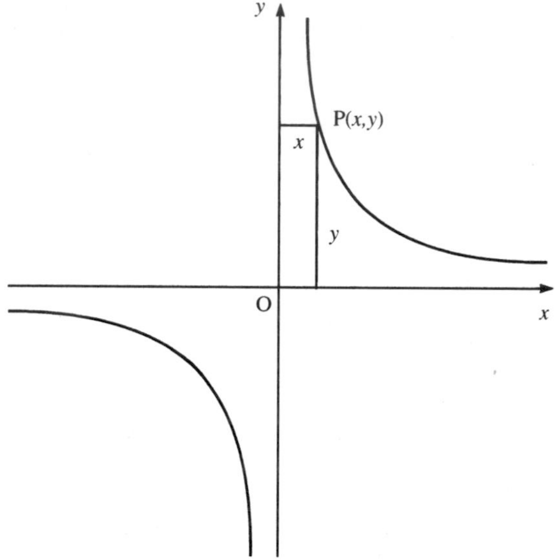

Figure 5.12

The lines through P parallel to the asymptotes have lengths x and y in this case, so the result of the example on page 105 becomes $xy = \frac{1}{4}(a^2 + a^2) = \frac{1}{2}a^2$. Putting $c^2 = \frac{1}{2}a^2$ gives the equation of a rectangular hyperbola referred to its asymptotes as co-ordinate axes in the very simple form $xy = c^2$. The particular case $c = 1$ shows that the familiar curve of reciprocals $y = \dfrac{1}{x}$ is a rectangular hyperbola.

The curve $xy = c^2$ has the simple parametric equations $x = ct$, $y = \dfrac{c}{t}$, where $t \neq 0$.

Activity

Check that the point $\left(ct, \dfrac{c}{t}\right)$ lies on $xy = c^2$ and that every point of the curve corresponds to one and only one non-zero value of t. Describe how $\left(ct, \dfrac{c}{t}\right)$ moves along the curve as t increases from $-\infty$ to ∞.

The gradient of the chord joining the points where $t = t_1$ and $t = t_2$ is

$$\frac{\left(\frac{c}{t_2} - \frac{c}{t_1}\right)}{ct_2 - ct_1} = \frac{\left(\frac{t_1 - t_2}{t_1 t_2}\right)}{t_2 - t_1} = -\frac{1}{t_1 t_2}.$$

Therefore the equation of the chord is $y - \frac{c}{t_1} = -\frac{1}{t_1 t_2}(x - ct_1)$

$$\Leftrightarrow \quad t_1 t_2 y - ct_2 = -x + ct_1$$

$$\Leftrightarrow \quad x + t_1 t_2 y = c(t_1 + t_2).$$

By letting $t_2 \to t_1$ we see that the equation of the tangent to $xy = c^2$ at the point where $t = t_1$ is $x + t_1^2 y = 2ct_1$.

EXAMPLE

The vertices of a triangle are on a rectangular hyperbola. Prove that the orthocentre of the triangle is also on this hyperbola.

Solution

Let the triangle be $P_1P_2P_3$, and let the perpendicular from P_3 to P_1P_2 meet the hyperbola again at P_4 (see the diagram below).

Let the parameter of each P_i be t_i ($i = 1, 2, 3, 4$). The gradient of P_1P_2 is $-\frac{1}{t_1 t_2}$ and the gradient of P_3P_4 is $-\frac{1}{t_3 t_4}$. Since these lines are perpendicular the product of their gradients is -1. Therefore $t_1 t_2 t_3 t_4 = -1$. But this condition is symmetrical in t_1, t_2, t_3, t_4 and so P_1P_4 is also perpendicular to P_2P_3. Therefore P_4 also lies on the perpendicular from P_1 to P_2P_3, and so P_4 is the orthocentre of triangle $P_1P_2P_3$.

Similarly each of P_1, P_2, P_3 is the orthocentre of the triangle formed by the other three points.

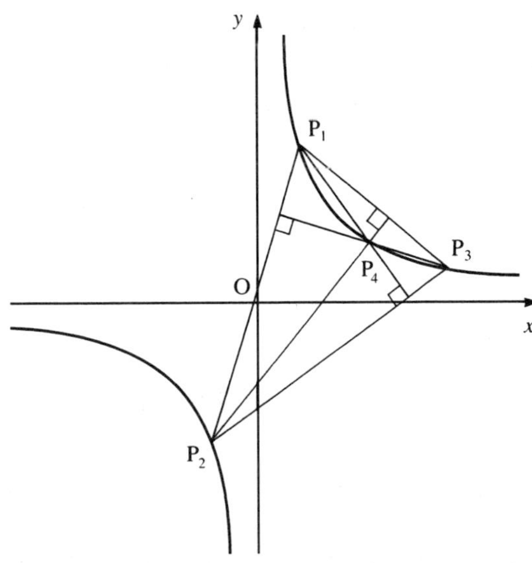

Exercise 5D

Throughout this exercise ***R*** is the rectangular hyperbola $xy = c^2$.

1. The tangent at a point P of ***R*** meets the co-ordinate axes at U and V. Prove that UV and OP make equal angles with the co-ordinate axes. Deduce that P is the midpoint of UV.
2. Find the co-ordinates of the foci and the equations of the directrices of ***R***.
3. Find the co-ordinates of the two points where the hyperbolas $x^2 - y^2 = 5$ and $xy = 6$ intersect. Prove that the tangents to the hyperbolas at these points form a rectangle.
4. The tangents at the points P and Q of ***R*** meet at T. Prove that the line joining T to the midpoint of PQ passes through the origin.
5. Sketch on a single diagram three members of each of the following families of rectangular hyperbolas:

   (A) those with equations $x^2 - y^2 = a^2$ for various a;

   (B) those with equations $xy = c^2$ for various c.

   Prove that every member of family (A) meets every member of family (B) at right angles, unless $a = c = 0$.
6. Find the equation of the normal to ***R*** at the point $P\left(ct, \frac{c}{t}\right)$. Prove that this normal meets the hyperbola again at $Q\left(-\frac{c}{t^3}, -ct^3\right)$.
   The circle with PQ as diameter meets ***R*** again at R. Prove that PR passes through the origin, and that the normal at R is parallel to PQ.
7. The midpoint of the chord joining $\left(ct, \frac{c}{t}\right)$ and $\left(cT, \frac{c}{T}\right)$ has co-ordinates (X, Y).

   Prove that $t + T = \frac{2X}{c}$ and $tT = \frac{X}{Y}$.

   A variable chord of ***R*** passes through the fixed point (h, k). Prove that the locus of the midpoint of the chord is another rectangular hyperbola, and give the equations of its asymptotes.
8. In this question P, Q, R, S are points on ***R*** with parameters p, q, r, s respectively.

   (i) Form a fourth degree equation in t by substituting $x = ct$, $y = c/t$ in

   $$x^2 + y^2 + 2gx + 2fy + k = 0.$$

   Deduce that if P, Q, R, S are concyclic (i.e. lie on a circle) then $pqrs = 1$.

   (ii) Prove the converse of the result in (i), i.e. $pqrs = 1 \Rightarrow$ P, Q, R, S are concyclic.

   (iii) Prove that if P, Q, R, S lie on a circle and PQ is a diameter of ***R*** then RS is a diameter of the circle.

   (iv) A variable circle touches ***R*** at a fixed point P and meets ***R*** again at variable points R and S. Prove that the chord RS has a fixed direction.

   (v) The normal to ***R*** at P meets the hyperbola again at P′, and Q′, R′, S′ are defined similarly. Prove that:

   P, Q, R, S are concyclic
   $\Leftrightarrow$ P′, Q′, R′, S′ are concyclic.

Focal distance and reflector properties

The last four sections and their exercises have explored a few of the many properties of particular conics (parabolas, ellipses or hyperbolas) which follow from the original focus–directrix definition, SP = ePM. Now we deal with two further general properties which have useful applications.

Focal distance properties

These concern the *central conics* (ellipses and hyperbolas), which have two foci and two directrices. In the case of the ellipse, with the usual notation (figure 5.13),

$$\text{SP} = e\text{PM} \qquad \text{and} \qquad \text{S'P} = e\text{PM'}$$

so that

$$\begin{aligned} \text{SP} + \text{S'P} &= e(\text{PM} + \text{PM'}) \\ &= e\text{MM'} \\ &= e \times \frac{2a}{e} \\ &= 2a. \end{aligned}$$

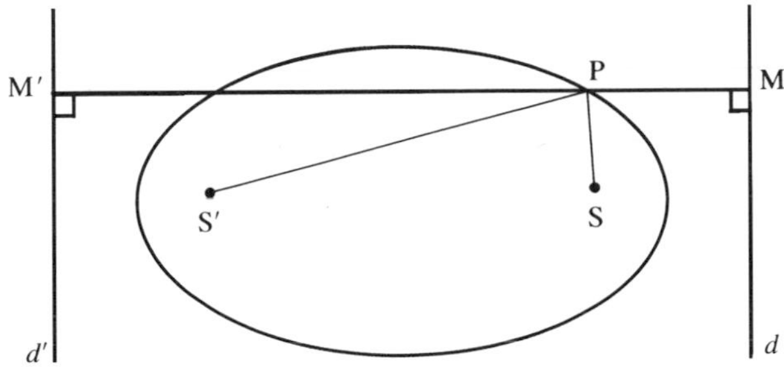

Figure 5.13

So the sum of the distances from the foci to any point of the ellipse is constant, and equals the length of the major axis.

This leads to the practical method of drawing an ellipse by fastening the ends of a piece of thread at S and S′ and then drawing with a pencil at P keeping the thread taut. (In practice it is easier to use a complete loop of thread PSS′.)

Activity

Draw some ellipses by this method. Notice the effect of

(i) changing the separation of S and S′ without changing the length of the loop;

(ii) changing the length of the loop without changing S and S′.

The corresponding property for the hyperbola is proved in a similar way, but now the length MM′ is the difference of the lengths PM and PM′, so the difference of the distances from the foci to any point of a hyperbola is constant, and equals the length of the transverse axis.

Activity

Prove this in detail, by drawing the diagram for the hyperbola corresponding to figure 5.13 and showing that $\text{SP} - \text{S'P} = 2a$ for points P on one branch, and $\text{S'P} - \text{SP} = 2a$ for points on the other branch.

This property of hyperbolas is the basis of many modern navigation systems. Two radio beacons S, S′ transmit simultaneous radio pulses, which arrive at a ship P at slightly different times (unless SP = S′P). By measuring this time lag (electronically) the difference SP – S′P can be calculated; the locus of points for which SP – S′P takes this value is one branch of a particular hyperbola with foci S, S′. By using the same procedure with pulses from S and a third beacon S″ the ship can also be located on one branch of a second hyperbola with foci S, S″ (figure 5.14) The intersection of these two branches gives the position of the ship, which can be found either from a specially drawn chart showing families of hyperbolas or by computer.

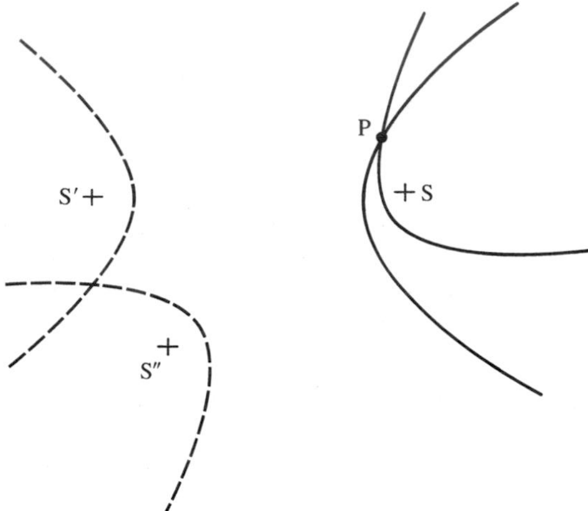

Figure 5.14

For Discussion

This method can be extended to fix positions in three dimensions (for example to track a robot exploring the sea bed). Discuss how to do this.

Reflector properties

The *focal distance properties* (SP = PM for the parabola, SP + S′P = $2a$ for the ellipse, $|\text{SP} - \text{S'P}| = 2a$ for the hyperbola) lead via a simple argument first used by Roberval in 1634 to the *reflector properties*. His idea was to consider a curve to be the path of a moving point P, so that the direction of the tangent at P is the direction of the velocity of P. Dealing first with the parabola, since PS and PM are always equal they must change at the same rate, so P must be moving away from S and from the directrix at equal speeds. Therefore if the velocity **v** of P makes angles ϕ and ψ with SP and MP respectively (figure 5.15) then $v \cos \phi = v \cos \psi$, and so $\phi = \psi$. This means that the tangent at P bisects angle SPM. (This has already been proved in Exercise 5A, Question 4.)

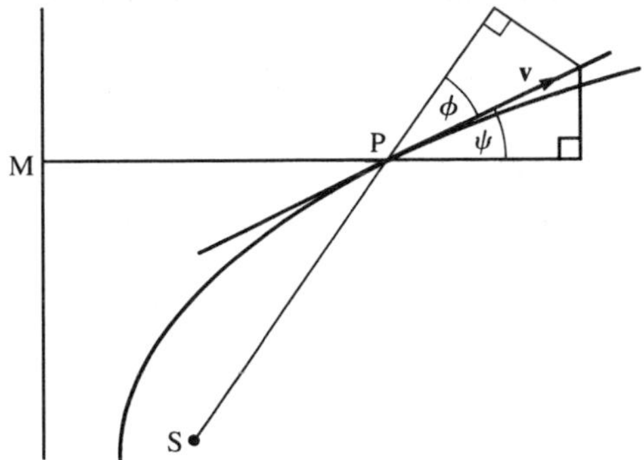

Figure 5.15

Activity

Prove that rays of light emitted from the focus and reflected by a parabolic mirror will produce a beam of light parallel to the axis, and that an incoming beam parallel to the axis will converge at the focus (= 'hearth' in Latin). Give some practical applications of this.

For the ellipse SP + S′P is constant, so the rate of increase of SP equals the rate of decrease of S′P. The velocity **v** of P has equal resolved parts in the directions of $\overrightarrow{SP}$ and $\overrightarrow{PS}$, i.e. $v \cos \phi = v \cos \psi$ (figure 5.16), and therefore $\phi = \psi$. This means that the tangent at P is equally inclined to SP and S′P.

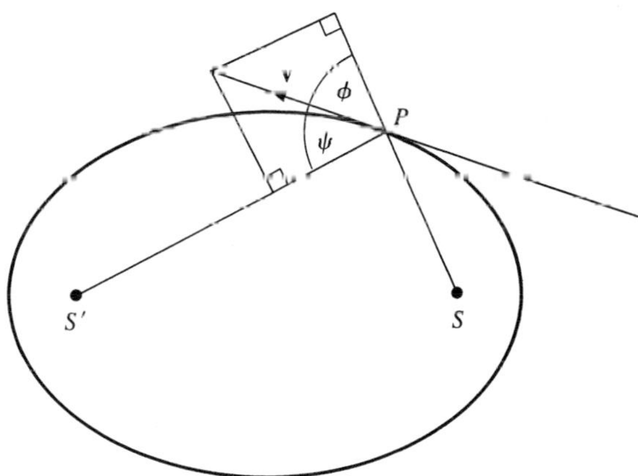

Figure 5.16

Activity

Prove by a similar method that the tangent at a point P of a hyperbola is equally inclined to the focal lines SP and S′P.

The reflector property of the ellipse is used in a modern medical device called a *lithotripter* (= 'stone crusher') (figure 5.17). This directs a powerful high frequency beam of sound waves at a kidney stone, so that the stone breaks into small fragments which pass from the body naturally. The sound source is at one focus of an elliptical mirror which is placed so that the stone is at the other focus. The beam spreads out from the source and then converges at the stone, which is thus the only part of the patient to receive the full effect of the beam. This prevents the sound waves from damaging the surrounding tissue.

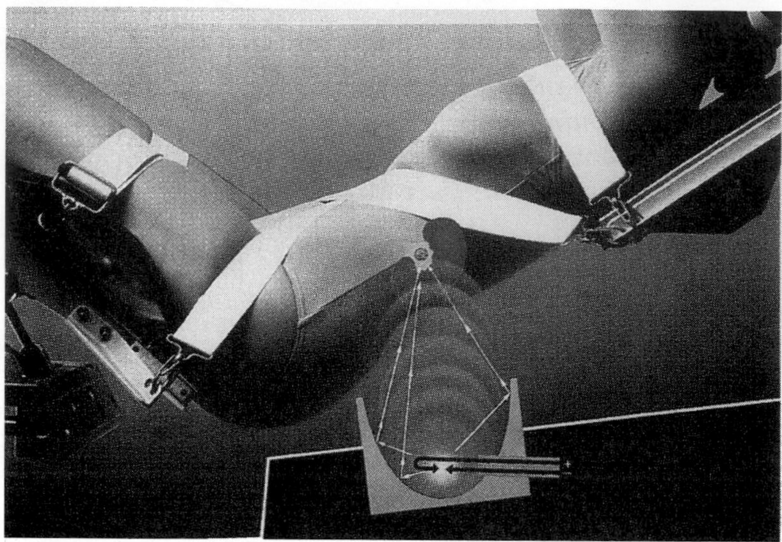

Figure 5.17

Exercise 5E

1. Check the property $SP + S'P = 2a$ for the ellipse of figure 5.16 by direct measurement from the diagram for several positions of P. Check similarly the corresponding property of the hyperbola from figure 5.11.

2. A variable circle passes through a fixed point S and touches a fixed straight line d. Prove that the locus of its centre is a parabola.

3. The diagram below shows a straight rod S′K of length ℓ which is pivoted at the point S′. A piece of string of length s ($< \ell$) has one end fastened at a fixed point S and the other at K. If the string is kept taut by a pencil P held against the rod, show that the locus of P is part of a hyperbola with transverse axis of length $\ell - s$.

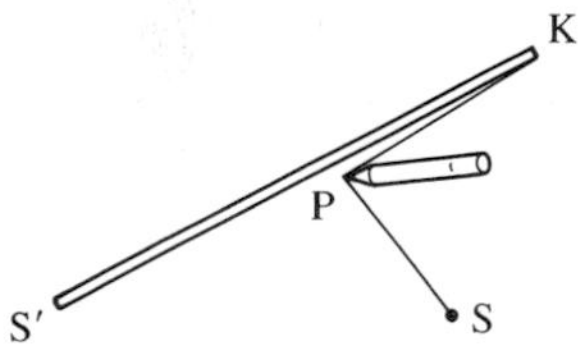

4. The variable circle C with centre P touches two fixed circles C_1 and C_2 with centres A_1 and A_2 respectively. Prove that if C_1 lies entirely outside C_2 then the locus of P is both branches of a hyperbola with foci A_1, A_2.

   Find the locus of P (i) if C_1 lies entirely inside C_2 ; (ii) if C_1 and C_2 intersect.

5. It is proposed to build a straight motorway connecting two towns A and B which are 200 km apart. In a rough model to investigate the effect this would have on the surrounding area it is assumed that traffic moves at 100 km/h on the motorway and at 50 km/h elsewhere, and that it is possible to drive directly to A or to B from any other point.

   Taking Cartesian co-ordinates so that A and B are (–100, 0) and (100, 0) respectively, find the set of points which would benefit from the motorway in the sense that to drive to A via B and the motorway would be quicker than driving directly to A.

6. (i) The diagram below shows a *crossed parallelogram linkage* in which A and B are fixed points, and BC, CD, DA are movable rods with AB = CD and BC = AD. The rods BC and AD cross at P; prove that the locus of P is an ellipse with foci A and B.

[**Hint:** consider the symmetry of the figure.]

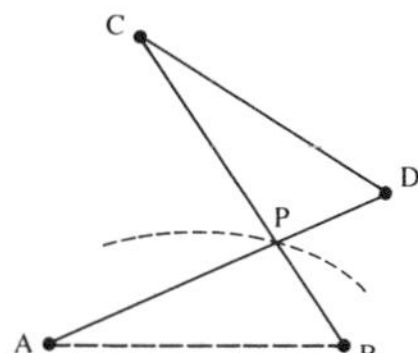

(ii) The next diagram shows a system of *elliptical gears,* consisting of the ellipse of (i) fixed to a shaft at A together with a congruent ellipse with foci C, D fixed to a shaft at D, so that A and D are now fixed, and both ellipses can rotate. The two gears are kept in contact by the rod BC, and have a common tangent at P.

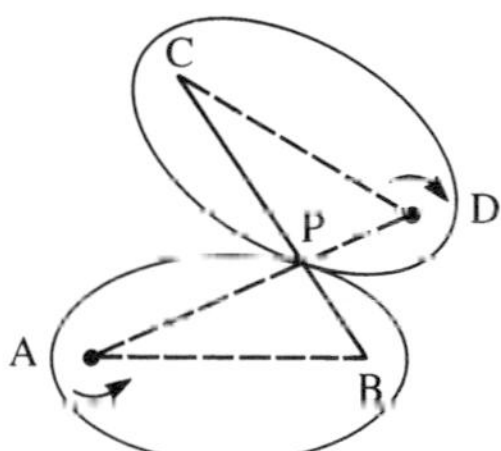

As the ellipse AB is made to turn it drives the ellipse CD, but the effective radii of the two gears, AP and DP, vary continuously, so that a constant speed input produces a variable speed output: this is used to provide a 'slow feed, quick return' mechanism in a number of machines. Prove that if the angular speed of ellipse AB is the constant ω then the angular speed of ellipse CD varies from $\dfrac{1-e}{1+e}\omega$ to $\dfrac{1+e}{1-e}\omega$.

7. Alternative proofs of the reflector properties.

(i) The point P of the ellipse $\dfrac{x^2}{a^2}+\dfrac{y^2}{b^2}=1$ has co-ordinates $(a\cos\theta, b\sin\theta)$.

Prove that the vector $\mathbf{t}=\begin{pmatrix}-a\sin\theta\\ b\cos\theta\end{pmatrix}$ is in the direction of the tangent at P.

(ii) Let $\overrightarrow{SP}=\mathbf{u}$ and $\overrightarrow{PS'}=\mathbf{v}$, where S and S′ are the foci. Prove that $|\mathbf{u}|=a(1-e\cos\theta)$ and $|\mathbf{v}|=a(1+e\cos\theta)$, and deduce that the vector $(1+e\cos\theta)\mathbf{u}+(1-e\cos\theta)\mathbf{v}$ is in the direction of the external bisector of angle SPS′.

(iii) Hence show that SP and S′P are equally inclined to the tangent at P.

(iv) Prove the reflector property of the hyperbola by a similar method.

8. (i) Given two fixed points S and S′ in a plane, use the focal distance properties to prove that through every other point of the plane there passes a unique ellipse with foci S, S′ and (except for points on the perpendicular bisector of SS′) a unique hyperbola with foci S, S′. This set of ellipses and hyperbolas is said to form a *confocal system.*

(ii) Use the reflector properties to prove that, in a confocal system, each ellipse meets each hyperbola *orthogonally,* i.e. that the tangents at the intersection are perpendicular. Sketch a confocal system to show this.

(iii) Describe the system of orthogonal curves which is obtained in the limit as $SS' \to 0$.

The conics as sections of a cone

The parabola, ellipse and hyperbola are called conics (or conic sections) because they were originally studied as plane sections of a right circular cone.

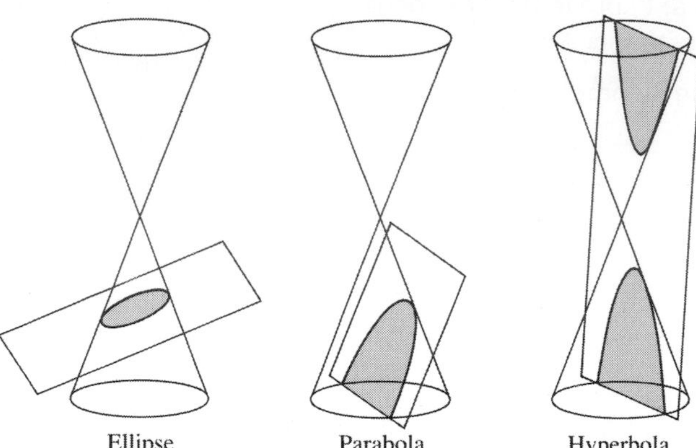

Figure 5.18

Figure 5.18 shows a double cone standing with its axis vertical. A horizontal plane not through the vertex cuts the cone in a circle. When the plane is tilted slightly the section is an ellipse. As the angle of tilt increases the section becomes more elongated until, when the plane is parallel to a generator of the cone (i.e. a straight line through the vertex in the surface of the cone), the section is a parabola. With further tilting, the plane cuts the other half of the cone too, and the section is a hyperbola. So the parabola is the borderline case, separating ellipses from hyperbolas.

For Discussion

Is it possible to obtain ellipses of all shapes and sizes as the sections of a single cone? Is the same true of hyperbolas?

HISTORICAL NOTE

In 1525 the German artist Albrecht Dürer (1471–1528) published a treatise on perspective and geometry which included drawing sections of a cone by using measurements from its plan and elevation. His ellipse (figure 5.19, page 117) is particularly interesting since he was convinced that the curve should widen in proportion with the widening of the cone, so he distorted the ellipse into an 'eierlinie' ('egg line' i.e. 'oval'). This error (by a master draughtsman with exceptionally acute perception) shows that it is necessary to prove that the sections of a cone are exactly the same as the curves produced from the focus–directrix definition with which we started on page 33. An elegant way of doing this, devised by Germinal Dandelin in 1822, is given in Questions 1 to 8 of the next exercise.

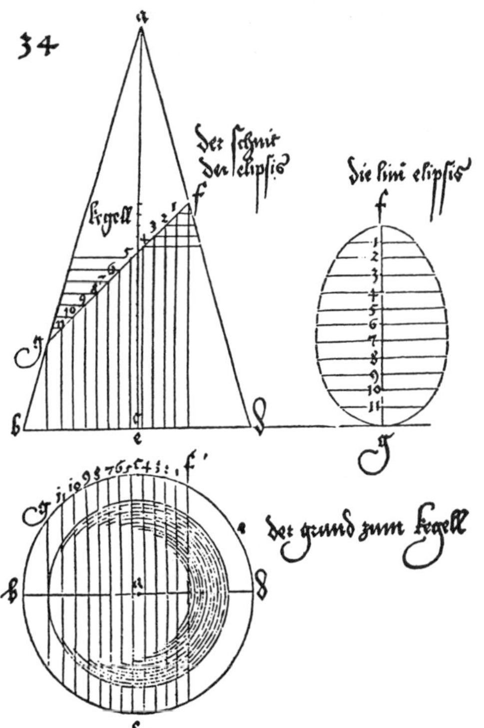

Figure 5.19

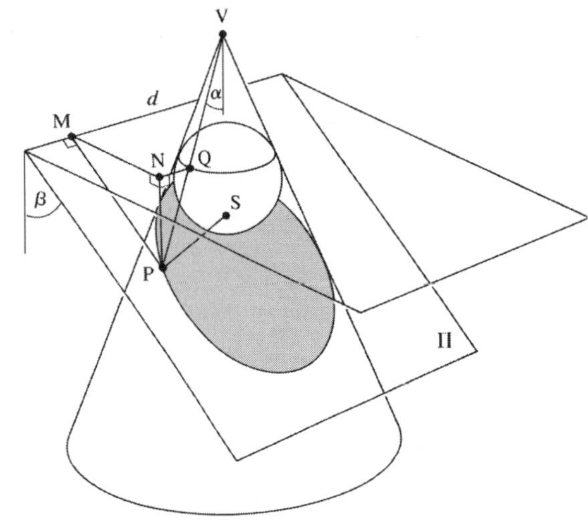

Figure 5.20

Exercise 5F

Most of this exercise refers to figure 5.20, which shows a plane Π cutting a cone with vertex V and a vertical axis. Any generator of the cone makes angle α with the vertical, and Π is inclined at angle β to the vertical. Between Π and V there is just one sphere which touches the cone and Π (to see this, imagine a tiny sphere touching the inside of the cone near V and then growing until it also touches Π). This *Dandelin sphere* touches Π at S and touches the cone in a horizontal circle. The horizontal plane containing this circle meets Π in the line d. From any point P of the conic section perpendiculars are drawn meeting d at M and meeting the horizontal plane at N. The generator PV meets the circle of contact at Q.

1 – 8. Prove the following properties of this figure.

1. $\mathrm{PS} = \mathrm{PQ}$.

2. $\angle\mathrm{NPQ} = \alpha$.

3. $\mathrm{PN} = \mathrm{PS}\cos\alpha$.

4. $\angle\mathrm{NPM} = \beta$.

5. $\mathrm{PN} = \mathrm{PM}\cos\beta$.

6. $\mathrm{PS} = \dfrac{\cos\beta}{\cos\alpha}\mathrm{PM}$.

7. P lies on the conic with focus S, directrix d, and eccentricity $\dfrac{\cos\beta}{\cos\alpha}$.

8. This conic is an ellipse if $\beta > \alpha$, a parabola if $\beta = \alpha$, and a hyperbola if $\beta < \alpha$.

Exercise 5F continued

9. Show that in the cases of the ellipse and hyperbola there is a second Dandelin sphere touching the cone and Π, and describe its position in each case. Explain why the point S′ where this touches Π is the other focus of the conic.

10. Prove the focal distance properties (SP + S′P = $2a$ for the ellipse, $|\text{SP} - \text{S'P}| = 2a$ for the hyperbola) directly from the Dandelin figure.

11. Show that the co-ordinates (x, y, z) of a point on the line joining the points (x_0, y_0, z_0) and (f, g, h) are given by $\begin{pmatrix} x \\ y \\ z \end{pmatrix} = \begin{pmatrix} f \\ g \\ h \end{pmatrix} + \lambda \begin{pmatrix} x_0 - f \\ y_0 - g \\ z_0 - h \end{pmatrix}$.

Deduce that the cone which passes through the ellipse $\frac{x^2}{a^2} + \frac{y^2}{b^2} = 1$, $z = 0$ and has its vertex at (f, g, h), $h \neq 0$, has the equation

$$\frac{(fz - hx)^2}{a^2} + \frac{(gz - hy)^2}{b^2} = (z - h)^2.$$

What is the significance of the condition $h \neq 0$?

Show that this cone meets the plane $x = 0$ in a circle if the vertex lies on a certain hyperbola in the plane $y = 0$. [MEI]

12. An ellipse has major axis AA′ and foci S, S′. Prove that the locus of viewing points from which this ellipse is seen as a circle is the hyperbola with transverse axis SS′ and foci A, A′ in the plane perpendicular to the plane of the ellipse.

Investigation

Projectile trajectories

(i) A particle is thrown from the origin O with initial velocity $\begin{pmatrix} u \\ v \end{pmatrix}$ referred to horizontal and vertical axes. Its position vector after time t is $\begin{pmatrix} x \\ y \end{pmatrix} = \begin{pmatrix} ut \\ vt - \frac{1}{2}gt^2 \end{pmatrix}$, where g is the acceleration due to gravity. Show that the Cartesian equation of its trajectory (path) can be written as $\left(x - \frac{uv}{g}\right)^2 = -\frac{2u^2}{g}\left(y - \frac{v^2}{2g}\right)$. Deduce that the trajectory is a parabola, and find the co-ordinates of its focus and the length of its semi-latus rectum.

(ii) Show that the directrix of this parabola is $y = \frac{V^2}{2g}$, where $V = \sqrt{u^2 + v^2}$ is the initial speed. Note that the position of the directrix is independent of the angle of projection.

(iii) Show that if the particle is thrown vertically upward it will just reach the directrix.

(iv) Now suppose that the particle is thrown with speed V from a point O on a sloping plane so as to hit the plane at a point B further up the line of greatest slope through O.

Explain the following construction for the possible trajectories (figure 5.21): Draw the directrix at height $\frac{V^2}{2g}$ above O, and construct circles with centres O and B to touch this. Then the points S_1 and S_2 where these circles intersect are the foci of the two possible parabolic trajectories.

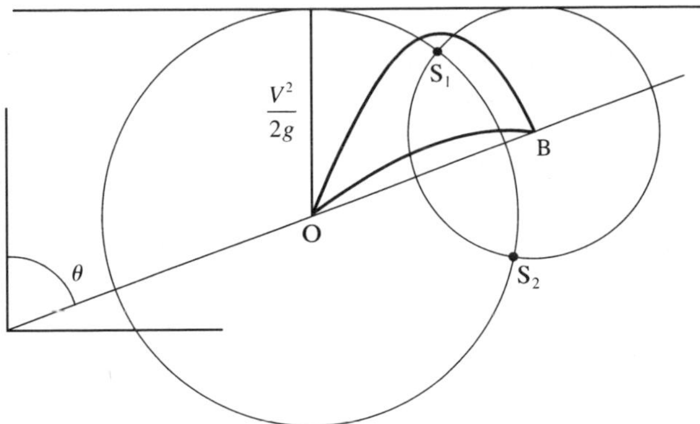

Figure 5.21

(v) By considering what happens as B moves up the slope, prove that

(a) the greatest range up the slope occurs when the focus of the trajectory lies on the slope;

(b) the direction of projection then bisects the angle between OB and the vertical;

(c) the direction of motion at B is then perpendicular to the direction of motion at O.

(vi) Show that the greatest range r up the slope is given by $r = \dfrac{V^2}{g(1+\cos\theta)}$,

where θ is the angle between the slope and the vertical.

(vii) By letting θ vary show that the set of points which the particle can hit when thrown from O with initial speed V is bounded by a parabola (called the *parabola of safety*). Describe the corresponding set of points in three dimensions.

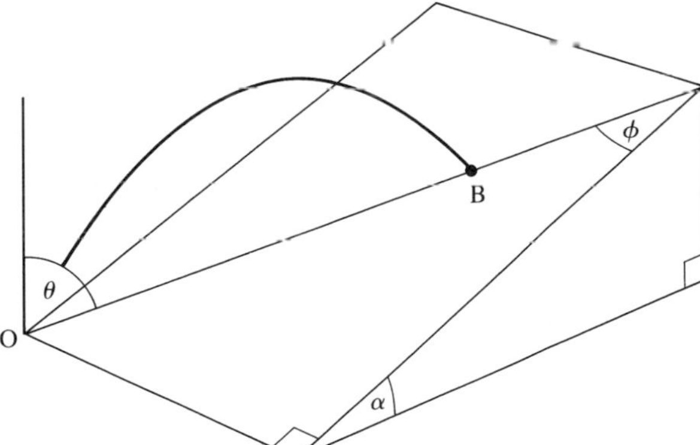

Figure 5.22

(viii) Finally suppose that the particle is thrown from a point O on a plane which is inclined at an angle α to the horizontal, hitting the plane at B, where OB makes an angle ϕ with the line of greatest slope OA and an angle θ with the vertical (figure 5.22).

Show that $\cos\theta = \sin\alpha\cos\phi$. Deduce that the set of points on the plane which the particle can hit is bounded by an ellipse with one focus at O and eccentricity $\sin\alpha$.

K E Y   P O I N T S

•

	Ellipse	Parabola	Hyperbola	Rectangular hyperbola
Standard form	$\frac{x^2}{a^2}+\frac{y^2}{b^2}=1$	$y^2=4ax$	$\frac{x^2}{a^2}-\frac{y^2}{b^2}=1$	$xy=c^2$
Parametric form	$(a\cos\theta, b\sin\theta)$	$(at^2, 2at)$	$(a\sec\theta, b\tan\theta)$	$\left(ct, \frac{c}{t}\right)$
Eccentricity	$e<1$ $b^2=a^2(1-e^2)$	$e=1$	$e>1$ $b^2=a^2(e^2-1)$	$e=\sqrt{2}$
Foci	$(\pm ae, 0)$	$(a, 0)$	$(\pm ae, 0)$	$(\pm\sqrt{2}c, \pm\sqrt{2}c)$
Directrices	$x=\pm\frac{a}{e}$	$x=-a$	$x=\pm\frac{a}{e}$	$x+y=\pm\sqrt{2}c$
Asymptotes	none	none	$\frac{x}{a}=\pm\frac{y}{b}$	$x=0, y=0$

Any of these conics can be expressed in polar co-ordinates (with the origin as the focus) as : $\frac{\ell}{r}=1+e\cos\theta$

where ℓ is the length of the semi-latus rectum.

- Focal distance properties: SP + SP′ = $2a$ for ellipse, |SP – SP′| = $2a$ for hyperbola.
- Reflector properties:

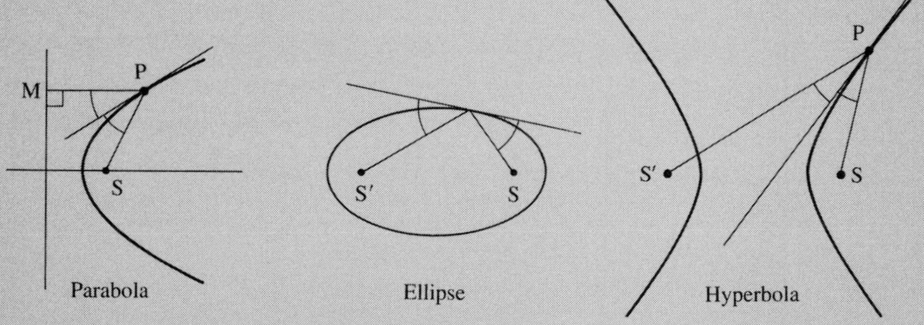

- The conics may be obtained as plane sections of a circular cone. If the cone has a vertical axis and semi-vertical angle α, and the cutting plane is inclined to the vertical at angle β then the eccentricity of the conic section is $\frac{\cos\beta}{\cos\alpha}$.

6 Hyperbolic functions

Out, hyperbolical fiend!

Shakespeare

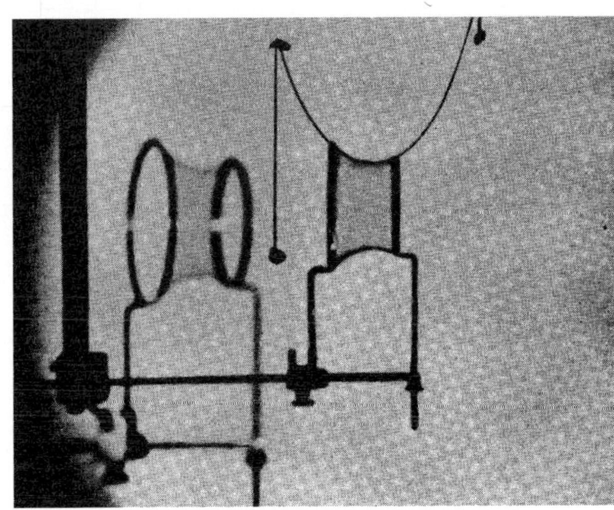

The soap bubble and the hanging chain both form the same curve – a catenary.

Hyperbolic functions

The cosine and sine functions are called *circular functions*, since the parametric equations $x = \cos\theta$, $y = \sin\theta$ give the circle $x^2 + y^2 = 1$. This equation can be rearranged to give $y = \pm\sqrt{1-x^2}$, which is why the inverse circular functions are useful in finding integrals involving $\sqrt{1-x^2}$ (and, likewise, $\sqrt{a^2-x^2}$, as on page 52). In the eighteenth century several mathematicians investigated integrals involving $\sqrt{x^2-1}$ in a similar way, noticing that if $y = \sqrt{x^2-1}$ then $x^2 - y^2 = 1$ which is the equation of a hyperbola (see page 103).

Now $\quad x^2 - y^2 = 1 \Leftrightarrow (x+y)(x-y) = 1$

so that if $\quad x + y = p$

then $\quad x - y = \dfrac{1}{p}$

from which $\quad x = \dfrac{1}{2}\left(p + \dfrac{1}{p}\right)$ and $y = \dfrac{1}{2}\left(p - \dfrac{1}{p}\right)$.

These are parametric equations for the hyperbola $x^2 - y^2 = 1$ in terms of the parameter p. (Compare Question 11 of Exercise 5C.)

These equations turn out to be particularly useful in the case when $p = \mathrm{e}^u$, so that $\frac{1}{p} = \mathrm{e}^{-u}$.

Then $\quad x = \frac{1}{2}(\mathrm{e}^u + \mathrm{e}^{-u})$ and $y = \frac{1}{2}(\mathrm{e}^u - \mathrm{e}^{-u})$.

By analogy with the circular functions these are called the *hyperbolic cosine* and *hyperbolic sine* functions respectively (names introduced by J H Lambert in 1768). These are abbreviated to *cosh* and *sinh* (pronounced 'shine' or 'sine–ch' or 'sinch' or 'sinsh' – take your pick!), so that

$$\cosh u = \frac{1}{2}(\mathrm{e}^u + \mathrm{e}^{-u}) \quad \text{and} \quad \sinh u = \frac{1}{2}(\mathrm{e}^u - \mathrm{e}^{-u}).$$

Activity

Prove that $\cosh(-u) = \cosh u$ and that $\sinh(-u) = -\sinh u$ (i.e. that cosh and sinh are respectively even and odd functions). What does this tell you about the symmetries of the graphs of these functions?

The graphs of these hyperbolic functions are easy to sketch. Since $\cosh u = \frac{1}{2}(\mathrm{e}^u + \mathrm{e}^{-u})$ the graph of $v = \cosh u$ lies midway between the graphs of $v = \mathrm{e}^u$ and $v = \mathrm{e}^{-u}$, as shown in figure 6.1. Note that $v = \cosh u$ has a minimum point at (0, 1).

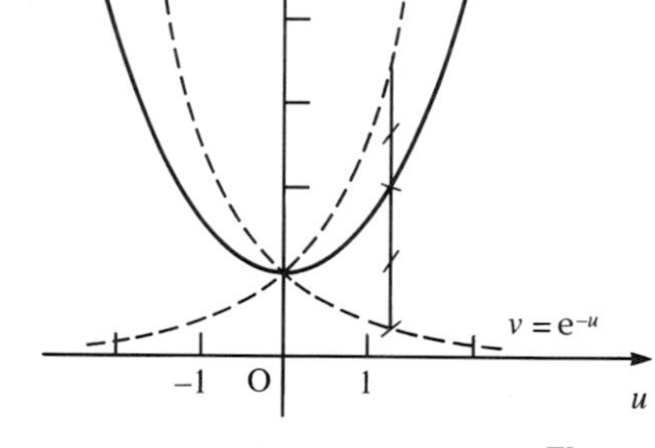

Figure 6.1

Similarly, the graph of $v = \sinh u$ is midway between the graphs of $v = \mathrm{e}^u$ and $v = -\mathrm{e}^{-u}$ (figure 6.2). It passes through the origin where it has a point of inflection.

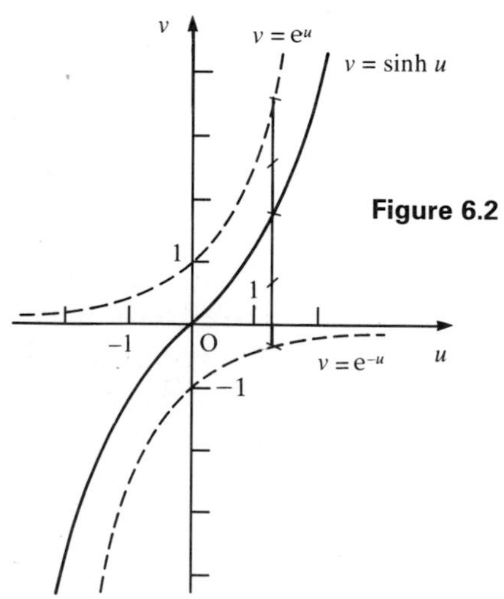

Figure 6.2

These graphs are nothing like the sine or cosine wave graphs, but the definitions of the hyperbolic functions are very similar to the results linking the circular functions with $e^{j\theta}$ which were given on page 84. Compare

$$\cosh u = \frac{1}{2}(e^{u} + e^{-u}) \qquad \text{and} \qquad \cos\theta = \frac{1}{2}(e^{j\theta} + e^{-j\theta})$$

$$\sinh u = \frac{1}{2}(e^{u} - e^{-u}) \qquad \text{and} \qquad \sin\theta = \frac{1}{2}(e^{j\theta} + e^{-j\theta}).$$

Many other similarities follow from this. For example, starting with the definitions and differentiating,

$$\frac{d}{du}(\cosh u) = \frac{1}{2}\left(e^{u} - e^{-u}\right) = \sinh u \qquad \text{and} \qquad \frac{d}{du}(\sinh u) = \frac{1}{2}\left(e^{u} + e^{-u}\right) = \cosh u.$$

Also, since $\quad \cosh^2 u = \tfrac{1}{4}(e^{u} + e^{-u})^2 = \tfrac{1}{4}(e^{2u} + 2 + e^{-2u})$

and $\quad \sinh^2 u = \tfrac{1}{4}(e^{u} + e^{-u})^2 = \tfrac{1}{4}(e^{2u} - 2 + e^{-2u})$

by subtracting $\cosh^2 u - \sinh^2 u = 1$

An important result, but not surprising since it gets us back to $x^2 - y^2 = 1$.

and by adding $\cosh^2 u + \sinh^2 u = \tfrac{1}{2}(e^{2u} + e^{-2u}) = \cosh 2u$.

Activity

Using $\cosh^2 u - \sinh^2 u = 1$ and $\cosh 2u = \cosh^2 u + \sinh^2 u$, write down two further versions of $\cosh 2u$. Compare all three formulae for $\cosh 2u$ with the corresponding formulae for $\cos 2\theta$.

Activity

Use the definitions of $\sinh u$ and $\cosh u$ to prove that

(i) $\sinh 2u = 2\sinh u \cosh u$;

(ii) $\sinh(u + v) = \sinh u \cosh v + \cosh u \sinh v$;

(iii) $\cosh(u + v) = \cosh u \cosh v + \sinh u \sinh v$.

[**Hint:** start with the right hand sides.]

The only difference between these identities and the corresponding ones for the circular functions is that the sign is reversed whenever a product of two sines is replaced by the product of two sinhs. This is called Osborn's Rule: it arises because of the factor j in the denominator of $\sin\theta$ as defined above.

EXAMPLE Solve the equation $\cosh u = 2\sinh u - 1$.

Solution

It is simplest to work from the definitions.

$$\begin{aligned}
\cosh u = 2\sinh u - 1 \quad &\Leftrightarrow \quad \tfrac{1}{2}(e^u + e^{-u}) = e^u - e^{-u} - 1 \\
&\Leftrightarrow \quad e^u - 3e^{-u} - 2 = 0 \\
&\Leftrightarrow \quad (e^u)^2 - 2e^u - 3 = 0 \\
&\Leftrightarrow \quad (e^u - 3)(e^u + 1) = 0 \\
&\Leftrightarrow \quad e^u = 3 \qquad \text{(since } e^u \text{ cannot be negative)} \\
&\Leftrightarrow \quad u = \ln 3.
\end{aligned}$$

Exercise 6A

1. Prove that

$$\cosh A + \cosh B = 2\cosh\frac{A+B}{2}\cosh\frac{A-B}{2}.$$

Write down the corresponding results for $\cosh A - \cosh B$ and for $\sinh A \pm \sinh B$, and prove one of these.

2. Given that $\sin 3\theta = 3\sin\theta - 4\sin^3\theta$ and $\cos 3\theta = 4\cos^3\theta - 3\cos\theta$, write down expressions for $\sinh 3u$ in terms of $\sinh u$ and $\cosh 3u$ in terms of $\cosh u$.

3. (i) Find all the real solutions of these equations.

(a) $\cosh x + 2\sinh x = -1$

(b) $10\cosh x - 2\sinh x = 11$

(c) $7\cosh x + 4\sinh x = 3$

(ii) Find conditions on a, b, c which are necessary and sufficient to ensure that the equation $a\cosh x + b\sinh x = c$ has two distinct real roots.

4. Given that $\sinh x + \sinh y = \dfrac{25}{12}$

and $\cosh x - \cosh y = \dfrac{5}{12}$,

show that $2e^x = 5 + 2e^{-y}$ and $3e^{-x} = -5 + 3e^y$.

Hence find the real values of x and y.

5. The figure on the right represents a cable hanging between two points A and B, where AB is horizontal. The lowest point of the cable, O, is taken as the origin of co-ordinates as shown.

If the cable is flexible and has uniform density then the curve in which it hangs is called a *catenary*. In 1691 John Bernoulli (responding to a challenge set by his brother James) proved that the equation of the catenary is $y = c(\cosh(x/c) - 1)$, where c is a constant [see *Pure Mathematics 6*, Chapter 4].

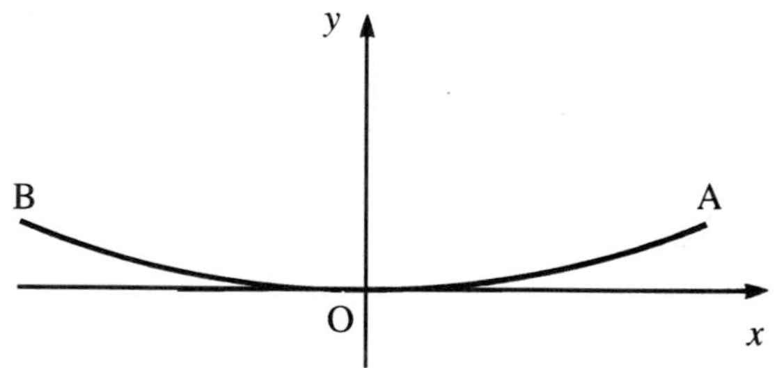

If for a particular cable $c = 20$ m and AB = 16 m, find the sag of the cable, i.e. the distance of O below AB, and the angle that the tangent at A makes with the horizontal.

6. P is any point on the curve $y = c\cosh(x/c)$, M is the foot of the perpendicular from P to the x axis, and Q is the foot of the perpendicular from M to the tangent of the curve at P. Prove that

(i) $MQ = c$;

(ii) the product of the y co-ordinates of P and Q is c^2.

7. Differentiate each of the following with respect to x.

(i) $\sinh 4x$

(ii) $\cosh(x^2)$

(iii) $\cosh^2 x$

(iv) $\cos x \sinh x$

(v) $\sinh(\ln x)$

(vi) $e^{5x}\sinh 5x$

(vii) $(1 + x)^3\cosh^3 3x$

(viii) $\ln(\cosh x + \sinh x)$

8. Express $\cosh^2 x$ and $\sinh^2 x$ in terms of $\cosh 2x$.

Hence find $\int \cosh^2 x\,dx$ and $\int \sinh^2 x\,dx$.

9. Integrate each of the following with respect to x.

(i) $\sinh 3x$

(i) $x\cosh(1 + x^2)$

(iii) $x\sinh x$

(iv) $\cos h^3 x$

(v) $x \sinh^2 x$

(vi) $e^{4x}\cosh 5x$

(vii) $\cosh^2 x \sinh^3 x$

(viii) $\cosh 6x \sinh 8x$

10. Use the Maclaurin series for e^x to find the Maclaurin series for $\cosh x$ and $\sinh x$, and compare them with the series for $\cos x$ and $\sin x$.

For what values of x are these series valid?

How many terms of the series are needed to give $\cosh 4$ correct to three significant figures?

11. Prove that $\cosh x > x$ for all x. Prove that the point on the curve $y = \cosh x$ which is closest to the line $y = x$ has co-ordinates $(\ln(1 + \sqrt{2}), \sqrt{2})$.

12. Prove that $(\cosh x + \sinh x)^n = \cosh nx + \sinh nx$ for all integers n. State and prove the corresponding result for $(\cosh x - \sinh x)^n$. Deduce expressions for $\cosh 5x$ in terms of $\cosh x$ and for $\sinh 5x$ in terms of $\sinh x$.

Other hyperbolic functions

The four remaining hyperbolic functions are defined in a similar way to the corresponding circular functions:

$$\tanh x = \frac{\sinh x}{\cosh x}, \coth x = \frac{1}{\tanh x}, \operatorname{sech} x = \frac{1}{\cosh x}, \operatorname{cosech} x = \frac{1}{\sinh x}.$$

Activity

For each of these functions state any necessary restriction on the domain, give the range, and say whether the function is even or odd.

The most important of these is the tanh function (pronounced 'than' or 'tan–ch').

Let $y = \tanh x$. Then $\quad y = \dfrac{e^x - e^{-x}}{e^x + e^{-x}}$

$$= \frac{1 - e^{-2x}}{1 + e^{-2x}}$$

dividing top and bottom by e^x

so that $\quad y \to 1$ as $x \to \infty$.

By a similar method, $\quad y \to -1$ as $x \to -\infty$.

Using the quotient rule to differentiate $\dfrac{\sinh x}{\cosh x}$ gives

$$\frac{dy}{dx} = \frac{\cosh x \cosh x - \sinh x \sinh x}{\cosh^2 x} = \text{sech}^2 x, \quad \text{since } \cosh^2 x - \sinh^2 x = 1.$$

So the graph of $y = \tanh x$ always has a positive gradient not exceeding 1 (since $0 < \text{sech}\, x \leqslant 1$), has half-turn symmetry about the origin, and has asymptotes $y = \pm 1$ (see figure 6.3).

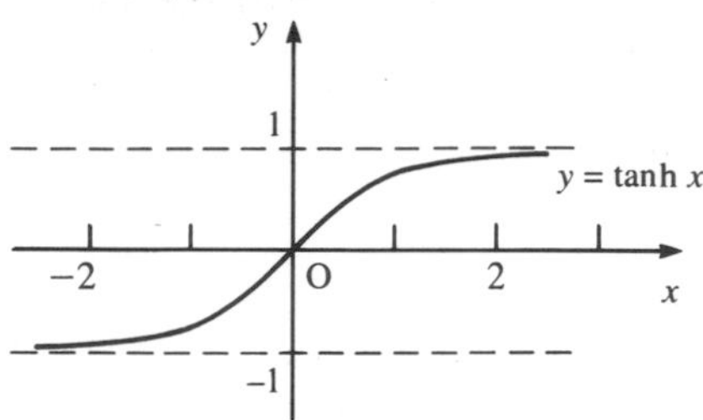

Figure 6.3

Exercise 6B

1. Sketch the graph of each of the following, giving the equations of any asymptotes.

(i) $y = \text{sech}\, x$

(ii) $y = \text{cosech}\, x$

(iii) $y = \coth x$

2. Prove that

(i) $1 - \tanh^2 x = \text{sech}^2 x$;

(ii) $\coth^2 x - 1 = \text{cosech}^2 x$;

(iii) $\tanh 2x = \dfrac{2\tanh x}{1 + \tanh^2 x}$

3. Find all the real solutions of these equations.

(i) $4\tanh x = \coth x$

(ii) $3\tanh x = 4(1 - \text{sech}\, x)$

(iii) $3\,\text{sech}^2 x + \tanh x = 3$

4. Find exact expressions for p and q, where $\sinh p = \text{sech}\, p$ and $\cosh q = \coth q$.

Arrange $\cosh x$, $\sinh x$, $\tanh x$, $\text{sech}\, x$, $\text{cosech}\, x$, $\coth x$ in ascending order of magnitude

(i) when $0 < x < p$;

(ii) when $p < x < q$.

5. If $-\pi/2 < x < \pi/2$ and k is any real constant, show that the equation $\sin x = \tanh k$ has just one solution, and prove that $\tan x = \sinh k$ and $\sec x = \cosh k$ for this value of x.

6. Prove that:

(i) $\alpha = \ln(\tan\beta) \Leftrightarrow \tanh\alpha = -\cos 2\beta$:

(ii) $\alpha = \ln\left(\tan\left(\dfrac{\pi}{4} + \dfrac{\beta}{2}\right)\right) \Leftrightarrow \tanh\alpha = \sin\beta$;

7. Differentiate each of the following with respect to x.

(i) $\text{sech}\, x$

(ii) $\text{cosech}\, x$

(iii) $\coth x$

(iv) $\ln(\tanh x)$

8. Integrate each of the following with respect to x.

(i) $\tanh x$

(ii) $\coth x$

(iii) $\text{sech}\, x$

(iv) $\text{cosech}\, x$

[**Hint:** for (iii) and (iv): use the substitution $u = e^x$.]

9. Find the first three non-zero terms of the Maclaurin series for $\tanh x$.

10. Prove that $\dfrac{x}{\sinh x} - \dfrac{\tanh x}{x}$ is positive for small values of x. Indicate what 'small' means in this context.

The inverse hyperbolic functions

The cosh function is a many-to-one function, since more than one value of x can yield the same value of y (e.g. $\cosh x_1 = \cosh(-x_1) = y_1$ in figure 6.4).

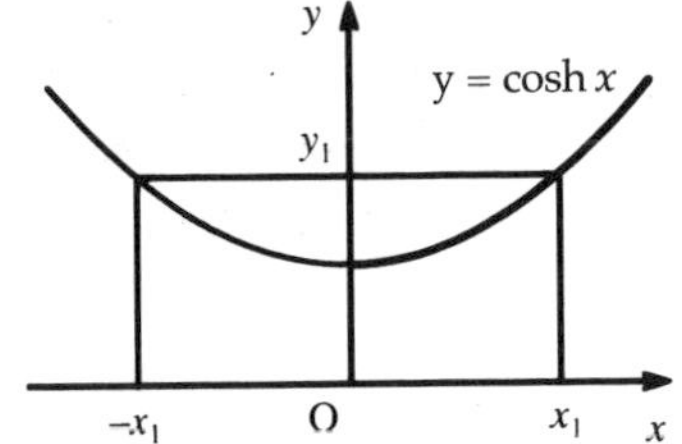

Figure 6.4

But if we restrict the domain to the non-negative real numbers, i.e. to $x \geqslant 0$, then the function is one-to-one, with the graph shown by the heavy line in figure 6.4. This restricted cosh function has an inverse function, which is denoted by *arcosh* (or sometimes $cosh^{-1}$), so that

$$v = \operatorname{arcosh} u \quad \Leftrightarrow \quad u = \cosh v \text{ and } v \geqslant 0.$$

The usual process of reflecting the graph of a function in the line $y = x$ to give the graph of its inverse function produces the graph of $y = \operatorname{arcosh} x$ shown in figure 6.5.

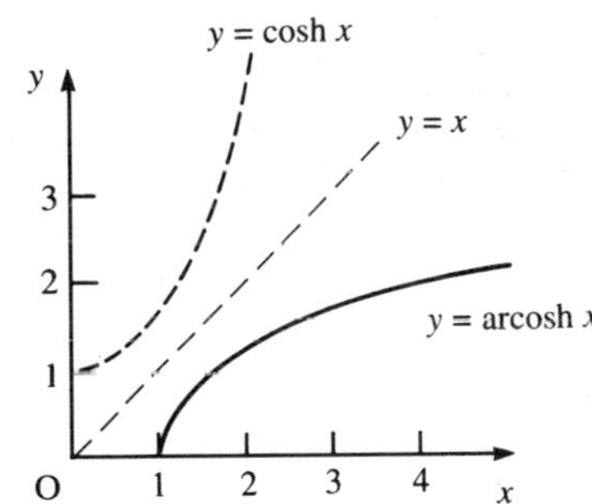

Figure 6.5

Since the sinh and tanh functions are already one-to-one there is no need for any similar restrictions in defining their inverse functions *arsinh* and *artanh* (or $sinh^{-1}$ and $tanh^{-1}$).

Thus $\quad v = \operatorname{arsinh} u \quad \Leftrightarrow \quad u = \sinh v$

$\qquad v = \operatorname{artanh} u \quad \Leftrightarrow \quad u = \tanh v.$

The graphs are shown in figure 6.6.

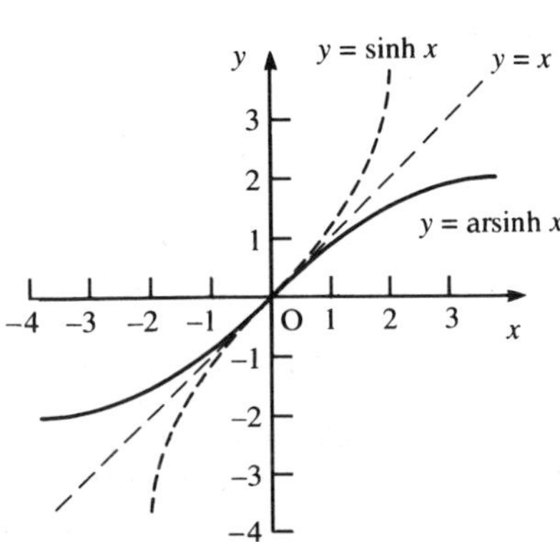

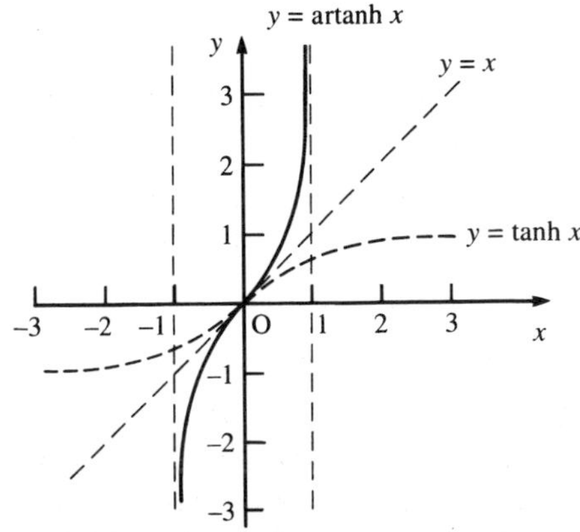

Figure 6.6

Activity

State the domain and range of each of these three inverse hyperbolic functions.

Just as the hyperbolic functions are defined using the exponential function, so their inverses can be put in terms of the natural logarithm function. The most straightforward to deal with is artanh x:

$$y = \text{artanh}\, x \quad \Leftrightarrow \quad x = \tanh y$$

$$= \frac{e^y - e^{-y}}{e^y + e^{-y}}$$

$$= \frac{e^{2y} - 1}{e^{2y} + 1} \quad \text{(multiplying top and bottom by } e^y\text{)}$$

$$\Leftrightarrow x(e^{2y} + 1) = e^{2y} - 1$$

$$\Leftrightarrow \quad e^{2y} = \frac{1+x}{1-x}$$

$$\Leftrightarrow \quad 2y = \ln\left(\frac{1+x}{1-x}\right).$$

Therefore $$\text{artanh}\, x = \frac{1}{2}\ln\left(\frac{1+x}{1-x}\right).$$

As you might expect, arcosh$x$ is a bit more complicated:

$$y = \text{arcosh}\, x \quad \Leftrightarrow \quad x = \cosh y$$

$$\Rightarrow \quad 2x = e^y + e^{-y}$$

$$\Rightarrow \quad (e^y)^2 - 2xe^y + 1 = 0$$

$$\Rightarrow \quad e^y = \frac{2x \pm \sqrt{4x^2 - 4}}{2} \quad \text{(using the quadratic equation formula)}$$

$$= x \pm \sqrt{x^2 - 1}$$

$$\Rightarrow \quad y = \ln(x + \sqrt{x^2 - 1}) \text{ or } \ln(x - \sqrt{x^2 - 1}) \ .$$

The sum of these two roots is

$$\ln\left[x + \sqrt{x^2 - 1})(x - \sqrt{x^2 - 1}\right] = \ln\left[x^2 - (x^2 - 1)\right] = \ln 1 = 0 \quad ,$$

so the second root is the negative of the first (as shown in figure 6.5).

Since arcosh $x \geqslant 0$ by definition, the root we want is the positive one.

Therefore $\quad \text{arcosh}\, x = \ln(x + \sqrt{x^2 - 1})$.

Activity

Use a similar method to prove that arsinh x = $\ln(x + \sqrt{x^2 + 1})$.

Explain why the root $\ln(x - \sqrt{x^2 + 1})$ **is rejected.**

The derivatives of arcosh x and arsinh x can be found by differentiating these logarithmic versions, but it is easier to work as follows.

$$y = \operatorname{arcosh} x \quad \Leftrightarrow \quad \cosh y = x$$

$$\Rightarrow \quad \sinh y \frac{dy}{dx} = 1$$

differentiating both sides with respect to x

$$\Rightarrow \quad \frac{dy}{dx} = \frac{1}{\sinh y}$$

$$= \frac{1}{\pm\sqrt{\cosh^2 y}}$$

using $\cosh^2 y - \sinh^2 y = 1$

$$= \frac{1}{\pm\sqrt{x^2 - 1}}.$$

Since the gradient of $y = \operatorname{arcosh} x$ is always positive we must take the positive square root, and therefore $\frac{d}{dx}(\operatorname{arcosh} x) = \frac{1}{\sqrt{x^2 - 1}}$.

This result is equivalent to the integral $\int \frac{1}{\sqrt{x^2 - 1}} dx = \operatorname{arcosh} x + c$,

from which it is easy to integrate related functions. For example, to find $\int \frac{1}{\sqrt{x^2 - a^2}} dx$, use the substitution $x = au$. Then $dx = a\,du$ and

$$\int \frac{1}{\sqrt{x^2 - a^2}} dx = \int \frac{1}{\sqrt{a^2u^2 - a^2}} a\,du = \int \frac{1}{a\sqrt{u^2 - 1}} a\,du = \int \frac{1}{\sqrt{u^2 - 1}} du$$

$$= \operatorname{arcosh} u + c$$

$$= \operatorname{arcosh} \frac{x}{a} + c.$$

Activity

Prove by similar methods that

(i) $\frac{d}{dx}(\operatorname{arsinh} x) = \frac{1}{\sqrt{x^2 + 1}}$; (ii) $\int \frac{1}{\sqrt{x^2 + a^2}} dx = \operatorname{arsinh} \frac{x}{a} + c$.

EXAMPLE

Find (i) $\int \frac{1}{\sqrt{9x^2 - 25}} dx$; (ii) $\int \sqrt{x^2 - 1}\, dx$;

(iii) $\int_1^5 \frac{1}{\sqrt{x^2 + 6x + 13}} dx$.

Solution

(a) $$\int \frac{1}{\sqrt{9x^2-25}}\,\mathrm{d}x = \frac{1}{3}\int \frac{1}{\sqrt{x^2-\frac{25}{9}}}\,\mathrm{d}x$$

$$= \frac{1}{3}\operatorname{arcosh}\frac{3x}{5}+c$$

(b) Let $x = \cosh u$ so that $\mathrm{d}x = \sinh u\,\mathrm{d}u$. Then

$$\int \sqrt{x^2-1}\,\mathrm{d}x = \int \sqrt{\cosh^2 u - 1}\,\sinh u\,\mathrm{d}u$$

$$= \int \sinh^2 u\,\mathrm{d}u$$

$$= \int \frac{1}{2}(\cosh 2u - 1)\mathrm{d}u$$

$$= \frac{1}{4}\sinh 2u - \frac{1}{2}u + c$$

$$= \frac{1}{2}\sinh u\cosh u - \frac{1}{2}u + c$$

$$= \frac{1}{2}x\sqrt{x^2-1} - \frac{1}{2}\operatorname{arcosh}x + c.$$

(c) $x^2 + 6x + 13 = x^2 + 6x + 9 + 4 = (x + 3)^2 + 4$. Therefore

$$\int_1^5 \frac{1}{\sqrt{x^2+6x+13}}\,\mathrm{d}x = \int_1^5 \frac{1}{\sqrt{(x+3)^2+4}}\,\mathrm{d}x$$

$$= \left[\operatorname{arsinh}\frac{x+3}{2}\right]_1^5$$

$$= \operatorname{arsinh}4 - \operatorname{arsinh}2$$

$$= \ln(4+\sqrt{17}) - \ln(2+\sqrt{5})$$

$$\approx 0.651.$$

Exercise 6C

1. Differentiate $\ln(x+\sqrt{x^2-1})$ with respect to x, and show that your answer simplifies to $\dfrac{1}{\sqrt{x^2-1}}$.

2. Prove that $\dfrac{\mathrm{d}}{\mathrm{d}x}(\operatorname{artanh}x) = \dfrac{1}{1-x^2}$. By using partial fractions and integrating, deduce from this the logarithmic form of $\operatorname{artanh}x$.

3. Sketch the graphs of the inverse functions $y = \operatorname{arsech}x$, $y = \operatorname{arcosech}x$, $y = \operatorname{arcoth}x$, giving the domain and range of each.

4. Differentiate each of the following with respect to x.
(i) $\operatorname{arsinh}3x$
(ii) $\operatorname{arcosh}(x^2)$
(iii) $\arctan(\sinh x)$
(iv) $\operatorname{artanh}(\sin x)$
(v) $\operatorname{arsech}x$

5. Integrate the following with respect to x.
(i) $\operatorname{arcosh}x$
(ii) $\operatorname{arsinh}x$
(iii) $\operatorname{artanh}x$.
[**Hint:** write $\operatorname{arcosh}x = 1 \times \operatorname{arcosh}x$ and integrate by parts.]

Exercise 6C continued

6. Integrate the following with respect to x.

(i) $\dfrac{1}{\sqrt{4+x^2}}$ (ii) $\dfrac{1}{\sqrt{x^2-9}}$

(iii) $\dfrac{1}{\sqrt{9-x^2}}$ (iv) $\dfrac{1}{\sqrt{36x^2+16}}$

(v) $\dfrac{1}{\sqrt{x^2-4x+8}}$ (vi) $\dfrac{1}{\sqrt{x^2+x}}$

(vii) $\dfrac{1}{\sqrt{9x^2+6x-8}}$ (viii) $\dfrac{x^2}{\sqrt{x^6-1}}$

7. Evaluate each of the following, correct to three significant figures.

(i) $\displaystyle\int_1^3 \frac{1}{\sqrt{x^2+4x+5}}\,dx$

(ii) $\displaystyle\int_{10}^{20} \frac{1}{\sqrt{4x^2+12x-40}}\,dx$.

8. The points $P_1(a\cos\theta, a\sin\theta)$ and $P_2(a\cosh\phi, a\sinh\phi)$ lie on the circle $x^2+y^2=a^2$ and the rectangular hyperbola $x^2-y^2=a^2$ respectively (see diagram above right).

Prove that area OAP_1 is proportional to θ and that area OAP_2 is proportional to ϕ, with the same constant of proportionality.

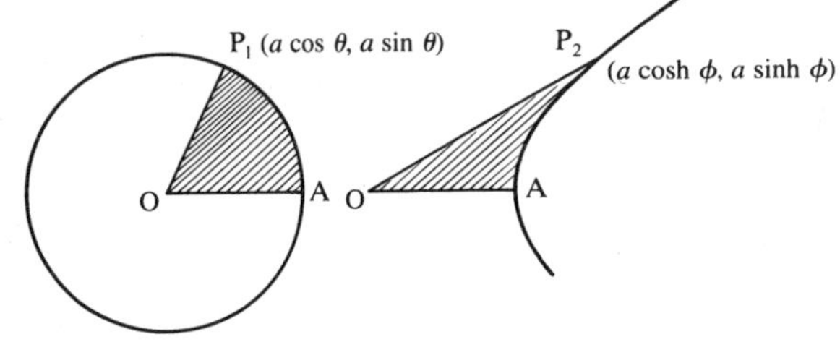

9. By substituting suitable circular or hyperbolic functions, find

(i) $\displaystyle\int\sqrt{a^2-x^2}\,dx$; (ii) $\displaystyle\int\sqrt{a^2+x^2}\,dx$;

(iii) $\displaystyle\int\sqrt{x^2-a^2}\,dx$.

10. Find the Maclaurin expansion of artanh x, stating the values of x for which it is valid. [**Hint:** use the logarithmic form, or use the result of Question 2].

11. In this question y_n stands for $\dfrac{d^ny}{dx^n}$ and a_n stands for the value of $\dfrac{d^ny}{dx^n}$ when $x=0$.

If $y = \text{arsinh}\,x$

(i) prove that $a_1=1, a_2=0$ and $(1+x^2)y_2+xy_1=0$;

(ii) prove by induction that $(1+x^2)y_{n+2}+(2n+1)xy_{n+1}+n^2y_n=0$ for $n \geqslant 0$,

and deduce that $a_{n+2}=-n^2a_n$.
Hence find the Maclaurin expansion of arsinh x.

Circular and hyperbolic functions of a complex variable

In the last section of Chapter 4 we defined e^z for a complex number z, and saw how this led to the following definitions of the circular functions:

$$\cos z = \frac{e^{jz}+e^{-jz}}{2} \quad \text{and} \quad \sin z = \frac{e^{jz}-e^{-jz}}{2j}.$$

It is natural to give similar definitions for the hyperbolic functions:

$$\cosh z = \frac{e^z+e^{-z}}{2} \quad \text{and} \quad \sinh z = \frac{e^z-e^{-z}}{2}.$$

Activity

Prove from these definitions that these four functions are periodic, with $\cos z$ and $\sin z$ having period 2π and $\cosh z$ and $\sinh z$ having period $2\pi j$.

Activity

Prove Euler's formulae: (a) $\cos jz = \cosh z$; (b) $\sin jz = j\sinh z$; (c) $\cosh jz = \cos z$; (d) $\sinh jz = j\sin z$.

The results of the last activity make it possible to find the real and imaginary parts of these trigonometric or hyperbolic functions. For example

$$\cos z = \cos(x+jy) = \cos x \cos jy - \sin x \sin jy$$
$$= \cos x \cosh y - j\sin x \sinh y.$$

This in turn enables you to solve problems which until now have seemed impossible, as in the following example.

EXAMPLE

Solve the equation $\cos z = 5$.

Solution

$$\cos(x+jy) = 5 \quad \Leftrightarrow \cos x \cosh y - j\sin x \sinh y = 5$$
$$\Leftrightarrow \cos x \cosh y = 5 \quad \text{and} \quad \sin x \sinh y = 0.$$
$$\text{Now } \sin x \sinh y = 0 \quad \Leftrightarrow \sin x = 0 \quad \text{or} \quad \sinh y = 0.$$

If $\sinh y = 0$ then $y = 0$, so $\cosh y = 1$ and $\cos x = 5$, which is impossible since x is real.

If $\sin x = 0$ then $x = n\pi$, so $\cos x = (-1)^n$ and $(-1)^n \cosh y = 5$.

This is impossible if n is odd, but if n is even it gives $\cosh y = 5$, so $y = \pm\,\text{arcosh}\,5 \approx \pm\,2.292$.

So the equation $\cos z = 5$ has infinitely many solutions, $z = 2k\pi \pm j\,\text{arcosh}\,5$. On the Argand diagram these give points spaced at intervals of 2π along the parallel lines $y \approx \pm\,2.292$ (figure 6.7).

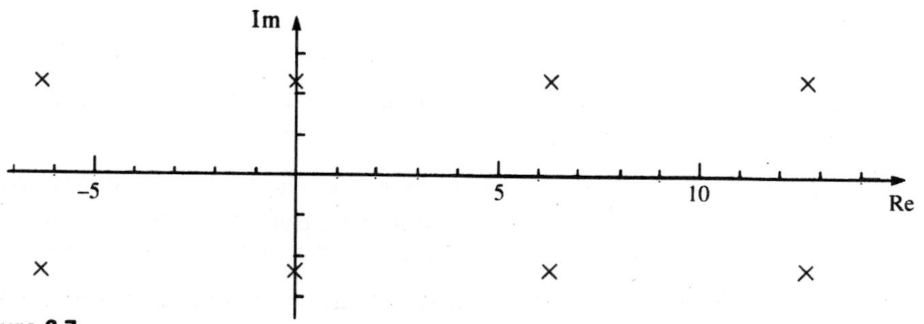

Figure 6.7

This brings you to the threshold of the theory of functions of a complex variable, first studied systematically by Augustin-Louis Cauchy (1789–1857), which has proved to have many applications in other areas of pure mathematics, physics and engineering. This most fertile branch of mathematics has been called the mathematical joy of the nineteenth century.

Exercise 6D

1. Find the real and imaginary parts of $\cos(2 + 3j)$ and of $\cosh(2 + 3j)$, giving your answers correct to three decimal places.

2. Find the real and imaginary parts of $\sinh(x+jy)$.

3. Prove that $\cosh(j\pi - \theta) = -\cosh\theta$ and $\sinh(j\pi/2 - \theta) = j\cosh\theta$.

4. Give definitions of $\tan z$ and $\tanh z$ for the complex number z. What can be deduced about the periodicity of these functions?

5. Given that $w = \sin(x+jy)$

(i) find $\text{Re}(w)$ and $\text{Im}(w)$;

(ii) prove that $|w|^2 = \frac{1}{2}(\cosh 2y - \cos 2x)$;

(iii) prove that $\tan(\arg w) = \cot x \tanh y$.

6. Find z if $\cosh z = -1$.

7. Prove that if $h > 1$ then:

$\sin z = h \Leftrightarrow z = (2k+\frac{1}{2})\pi \pm j\,\text{arcosh}\,h$.

State the corresponding result if $h < -1$.

8. By writing $\tan z$ in terms of e^{jz} prove that

$$\arctan x = \frac{1}{2j}\ln\left(\frac{1+jx}{1-jx}\right).$$

Prove the same result by using complex partial fractions to find $\int \frac{1}{1+x^2}\,dx$.

From this logarithmic form deduce the power series expansion of $\arctan x$.

9. Using horizontal x and y axes and a vertical h axis, describe and sketch the set of points (x, y, h) for which $\cos z = h$, where h is real and $z = x + jy$.

Investigation

The point representing the complex number z in an Argand diagram (called the z-plane) is mapped to the point representing the complex number w in a second Argand diagram (the w-plane), where $w = f(z)$. As z moves along a straight line $\text{Re}(z) = h$ in the z-plane so w moves along a curve in the w-plane. The set of such curves for different real values of h is called Family A. Similarly, Family B is the set of curves in the w-plane which are the images of the lines $\text{Im}(z) = k$ for different real values of k.

(i) Show that if $f(z) = e^z$ then Family A consists of circles with centre at the origin, and Family B consists of half-lines radiating from the origin.

(ii) Show that if $f(z) = \cosh z$ then Family A consists of ellipses, Family B consists of hyperbolas, and all these conics are confocal (see page 115).

(iii) Investigate Family A and Family B for other functions $f(z)$ (for example, z^2 or $\cos z$ or $1/z$). If the working becomes intractable, consider using computer power to plot the curves.

(iv) Investigate the angles at which curves of Family A meet curves of Family B.

KEY POINTS

- $\cosh x = \dfrac{e^x + e^{-x}}{2}$, $\sinh x = \dfrac{e^x - e^{-x}}{2}$, $\tanh x = \dfrac{\sinh x}{\cosh x} = \dfrac{e^{2x} - 1}{e^{2x} + 1}$,

  $\coth x = \dfrac{1}{\tanh x}$, $\operatorname{sech} x = \dfrac{1}{\cosh x}$, $\operatorname{cosech} x = \dfrac{1}{\sinh x}$.
- $\cosh^2 x - \sinh^2 x = 1$, $1 - \tanh^2 x = \operatorname{sech}^2 x$, $\coth^2 x - 1 = \operatorname{cosech}^2 x$.
- $\cosh(x \pm y) = \cosh x \cosh y \pm \sinh x \sinh y$,

  $\sinh(x \pm y) = \sinh x \cosh y \pm \cosh x \sinh y$,

  $\tanh(x \pm y) = \left(\dfrac{\tanh x \pm \tanh y}{1 \pm \tanh \pm \tanh y}\right)$.
- $\dfrac{d}{dx}(\cosh x) = \sinh x$, $\dfrac{d}{dx}(\sinh x) = \cosh x$, $\dfrac{d}{dx}(\tanh x) = \operatorname{sech}^2 x$,

  $\int \cosh x \, dx = \sinh x + c$, $\int \sinh x \, dx = \cosh x + c$, $\int \tanh x \, dx = \ln(\cosh x) + c$.
- $\operatorname{arcosh} x = \ln(x + \sqrt{x^2 - 1})$, $\operatorname{arsinh} x = \ln(x + \sqrt{x^2 - 1})$,

  $\operatorname{artanh} x = \frac{1}{2} ln\left(\dfrac{1+x}{1-x}\right)$.
- $\displaystyle\int \frac{1}{\sqrt{x^2 - a^2}} \, dx = \operatorname{arcosh} \frac{x}{a} + c$, $\displaystyle\int \frac{1}{\sqrt{x^2 + a^2}} \, dx = \operatorname{arsinh} \frac{x}{a} + c$.
- $\cosh x = \displaystyle\sum_{r=0}^{\infty} \frac{x^{2r}}{(2r)!}$ for all x, $\sinh x = \displaystyle\sum_{r=0}^{\infty} \frac{x^{2r+1}}{(2r+1)!}$ for all x,

  $\operatorname{artanh} x = \displaystyle\sum_{r=0}^{\infty} \frac{x^{2r+1}}{2r+1}$ for $-1 < x < 1$.
- $\cosh z = \dfrac{e^z + e^{-z}}{2}$, $\sinh z = \dfrac{e^z - e^{-z}}{2}$.
- $\cos jz = \cosh z$, $\sin jz = j \sinh z$, $\cosh jz = \cos z$, $\sinh jz = j \sin z$.

Answers

Exercise 1A

1. $x^3 + 4x^2 - 7x + 2 \equiv (x + 3)(x^2 + x - 10) + 32$ **2.** $x^4 + 6x^2 + 12 \equiv (x - 5)(x^3 + 5x^2 + 31x + 155) + 787$

3. (i) -183 (ii) 6 **5.** $-3, -7, 2$

6. (i) $2x^4 - 3x^3 + 5x^2 - 5x - 3 \equiv (x^2 + 2)(2x^2 - 3x + 1) + x - 5$

(ii) $4x^5 - 2x^4 - 2x^3 + x^2 - 3x + 2 \equiv (2x^2 - 3)(2x^3 - x^2 + 2x - 1) + 3x - 1$

(iii) $x^4 + 2x^3 - 5x^2 + 4x + 9 \equiv (x^2 - x + 3)(x^2 + 3x - 5) - 10x + 24$

(iv) $3x^4 - 8x^3 + 29x^2 + 21 \equiv (x^2 - x + 7)(3x^2 - 5x + 3) + 38x$

7. (i) When polynomial $P(x)$ is divided by $x - a$ the remainder is $P(a)$

(ii) -7 (iii) 6 (iv) -3 (v) $g(2) = f'(2) = -48$

8. $x^2 - 1,\ x + 2$ **9.** $-4, 2, 2$

10. (i) (a) $(x - a)(x + a);\ (x - a)(x^2 + ax + a^2)$

11. (i) (b) $(x + a)(x^2 - ax + a^2);\ (x + a)(x^4 - ax^3 + a^2x^2 - a^3x + a^4)$

12. $3(x - y)(y - z)(z - x)$

13. $P(x) \equiv 0 \Leftrightarrow a = b = c = 0;$ $P(x) = 0 \Leftrightarrow x = \dfrac{-b \pm \sqrt{b^2 - 4ac}}{2a}$, provided $a \neq 0$

14. $r = \dfrac{P(a) - P(b)}{a - b},\ s = \dfrac{aP(b) - bP(a)}{a - b}$ **16.** (ii) RS

Exercise 1B

1. $a = 1, b = 4, c = 4$

2. $a = 1, b = -3, c = 3, d = -1$

4. (i) There are infinitely many possible values of a, b, c: $a = 2 + c, b = -1 - 2c$

(ii) Unique values of a, b, c: $a = 3, b = -3, c = 1$

(iii) There are no such values of a, b, c

5. $a = b = \pm 2, c = 2$

6. (iii) $y = \dfrac{A(x-b)(x-c)(x-d)}{(a-b)(a-c)(a-d)} + \dfrac{B(x-c)(x-d)(x-a)}{(b-c)(b-d)(b-a)} + \dfrac{C(x-d)(x-a)(x-b)}{(c-d)(c-a)(c-b)} + \dfrac{D(x-a)(x-b)(x-c)}{(d-a)(d-b)(d-c)}$

7. (i) $y = 3x^2 - 5x + 3$ (ii) $y = x^3 - 3x^2 + 2x + 1$

8. $h(x) = \dfrac{9}{4}x^2 - 6x + \dfrac{23}{4}$

13. (ii) $M_2 = 3, M_3 = 7, M_5 = 31, M_7 = 127$, all prime; $M_{11} = 2047 = 23 \times 89$, composite

14. (iii) $S(x)$ cannot be factorised into linear factors over the integers if $S(x)$ is prime for $2n + 1$ distinct integer values of x

Exercise 1C

1. $x = 4$ (repeated), $x = 2$ **2.** $x = 6$ (repeated), $x = 3$

3. $x = -3$ (repeated), $x = \dfrac{1}{2}$ **4.** $x = \dfrac{1}{3}$ (repeated), $x = -5$

5. $x = -\dfrac{3}{2}$ (repeated), $x = 3$ **6.** $x = -1$ (occurs three times), $x = 2$

7. $x = -1$ (repeated), $x = -2$ (repeated)

8. $f(x) = 0$ and $f^{(n-1)}(x) = 0$ share a root, where $f^{(n-1)}(x) \equiv (n - 1)$th derivative of $f(x)$

Exercise 1D

1. (i) $-\frac{7}{2}, 3$ (ii) $\frac{1}{5}, -\frac{1}{5}$ (iii) $0, \frac{2}{7}$ (iv) $-\frac{24}{5}, 0$ (v) $-11, -4$ (vi) $-\frac{8}{3}, -2$

2. (i) $z^2 - 10z + 21 = 0$ (ii) $2z^2 + 19z + 45 = 0$ (iii) $z^2 - 5z = 0$
(iv) $z^2 - 6z + 9 = 0$ (v) $z^2 - 6z + 13 = 0$

3. (i) $2, 3, \frac{2}{3}, \frac{1}{3}$ (ii) $3z^2 - 2z + 1 = 0$ (iii) $cz^2 + bz + a = 0$

4. (i) $2z^2 - 5z - 9 = 0$ (ii) $2z^2 + 15z + 16 = 0$ (iii) $4z^2 - 61z + 81 = 0$ (iv) $18z^2 + 61z + 18 = 0$

5. $z^2 - 16z - 8 = 0$

6. (i) $az^2 + bkz + ck^2 = 0$ (ii) $az^2 + (b - 2ka)z + (k^2a - kb + c) = 0$

7. (i) Distinct negative (real) roots (ii) $\alpha = -\beta$ (iii) one root is 0
(iv) distinct real roots, one positive, the other negative

9. (i) $3x^2 - 6x - (11 + k) = 0$; 1; locus of M is the vertical line: $x = 1$
(ii) locus of M is the vertical line: $x = \frac{m-b}{2a}$

Exercise 1E

1. (i) $-\frac{3}{2}$ (ii) $-\frac{1}{2}$ (iii) $-\frac{7}{2}$ (iv) $\frac{13}{4}$ (v) $-\frac{129}{8}$ (vi) $\frac{497}{16}$ (vii) $\frac{1}{7}$ (viii) $\frac{3}{7}$ (ix) $\frac{45}{4}$ (x) $-\frac{45}{14}$

2. (i) $z^3 - 8z^2 - 4z + 24 = 0$ (ii) $z^3 - 10z^2 + 27z - 19 = 0$ (iii) $z^3 - 8z^2 + 15z + 1 = 0$

3. (i) $2, 5, 8$ (ii) $-\frac{2}{3}, \frac{2}{3}, 2$ (iii) $2 - 2\sqrt{3}, 2, 2 + 2\sqrt{3}$ (iv) $\frac{2}{3}, \frac{7}{6}, \frac{5}{3}$

4. 1.5, 2, 2.5; 23.5

5. −0.75, 0.25, 0.5

6. (i) $a^2z^3 + (2ac - b^2)z^2 + (c^2 - 2bd)z - d^2 = 0$
(ii) $d^2z^3 - (c^2 - 2bd)z^2 - (2ac - b^2)z - a^2 = 0$
(iii) $a^2dz^3 - a(2bd - c^2)z^2 + d(b^2 - 2ac)z + ad^2 = 0$

7. $ac^3 = db^3$; 0.5, 1.5, 4.5

8. (i) 5, −6, 4 (ii) 37 (iv) $z^3 + z^2 + 37z - 4 = 0$

9. $p = 7, q = 8, \alpha = -1$; or $p = q = \alpha = 0$

10. (i) All permutations of (1, 3, −5) (ii) all permutations of (−1, 3, 4)
(iii) all permutations of (2, −3, 3)

11. (i) All permutations of (5, −2, −1) (ii) all permutations of (0, −3, 5)
(iii) all permutations of $(1, 2 + j, 2 - j)$

12. $\frac{7}{3}, \frac{3}{7}, -2$ **13.** $\pm\sqrt{(-q)}, -p$

14. (i) $-\frac{15}{4}$ (ii) $\frac{3}{4}$ (iii) -12 (iv) $\frac{135}{8}$

Exercise 2A

1. Kite

2. (a) $(4, 11\pi/12)$ and $(4, -5\pi/12)$ (b) $(4, 7\pi/12)$ or $(4, -\pi/12)$
(c) $(4/\sqrt{3}, 3\pi/4)$ and $(4/\sqrt{3}, -\pi/4)$, or $(4, -3\pi/4)$ and $(4\sqrt{3}, 3\pi/4)$, or $(4, -3\pi/4)$ and $(4\sqrt{3}, -\pi/4)$

3. (ii) A(5.39, 0.38), B(8.71, 1.01), C(8.71, 1.64), D(5.39, 2.27)
(iii) B(4.64, 7.37), C(−0.58, 8.69), D(−3.45, 4.14)

4. (i) 4, (ii) $16 < r < 170, \theta = -27$, (iii) (a) $99 < r < 107, 153 < \theta < 171$,
(b) $16 < r < 99$ or $107 < r < 162, -81 < \theta < -63$, (c) $16 < r < 99$ or $107 < r < 162$,
$45 < \theta < 63$ or $162 < r < 170, 153 < \theta < 135$ or $99 < r < 107, -9 < \theta < 9$

Exercise 2B

1. $x^2 + y^2 - 8y = 0$ **2.**

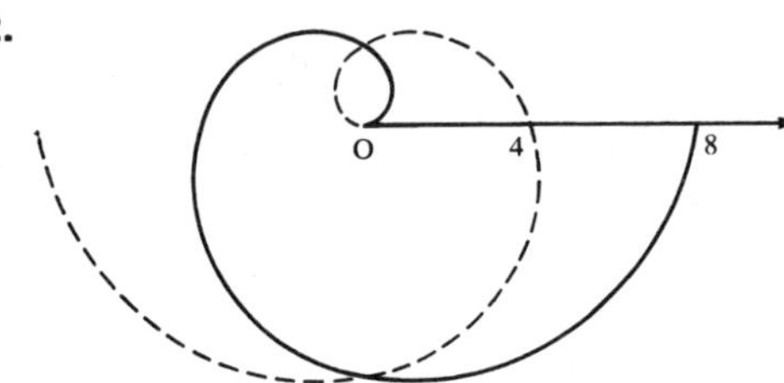

3. (i) Circle

(ii)

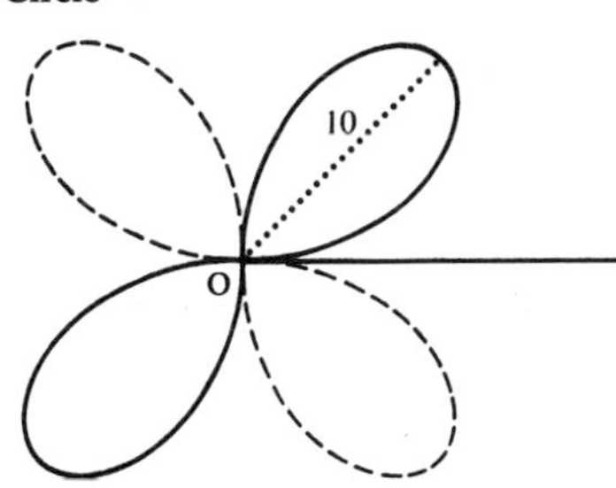

(iii)

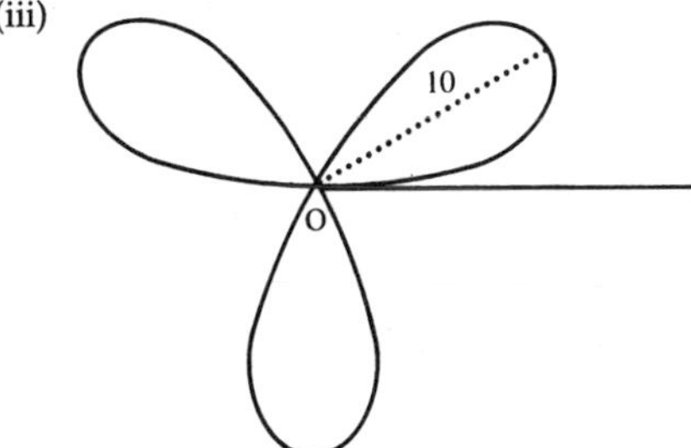

(iv) n repeated petals when n is odd, $2n$ petals when n is even

4.

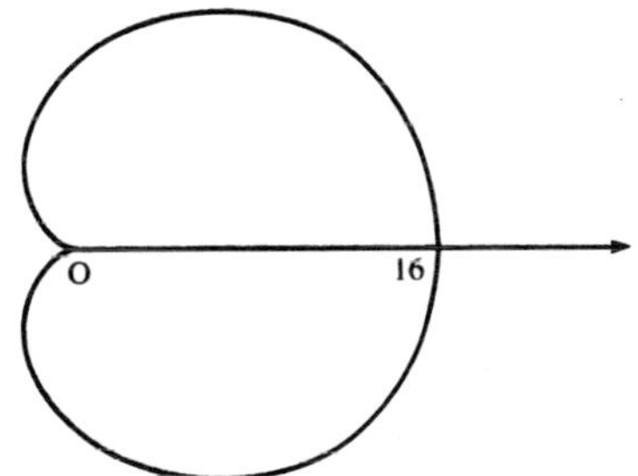

5. $x = a, y = b$ **6.** $x \cos \alpha + y \sin \alpha = p$

7. $(a, \pm\pi/3)$, $2r = a \sec\theta$

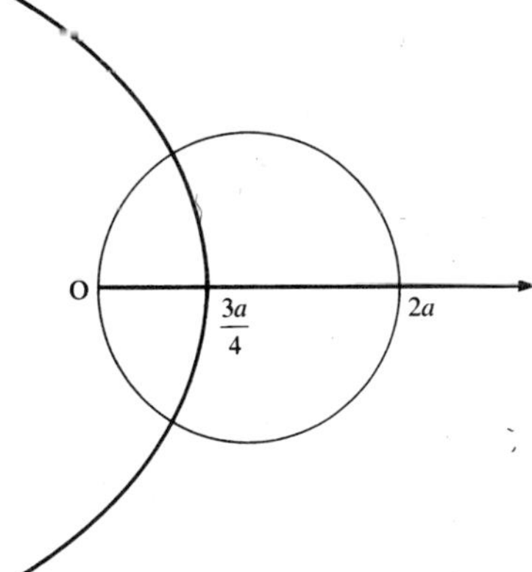

Activity page 32

(i) $2\pi + 3\sqrt{3}/2$ (ii) $\pi + 3\sqrt{3}$

Exercise 2C

1. It gives twice the area **2.** $64\pi/3$ **3.** $24\pi \pm 64$

4. $a^2/2$

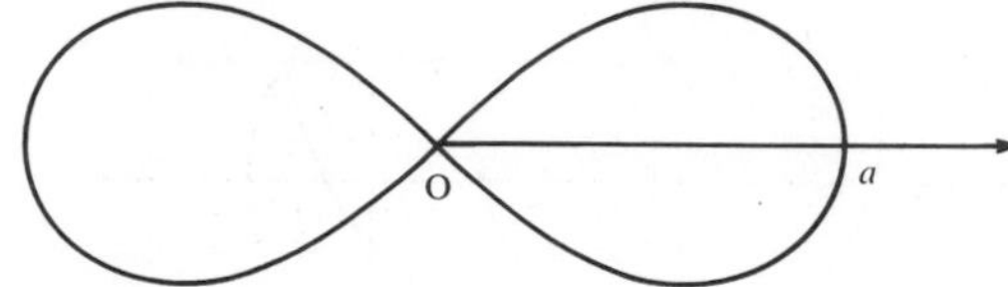

5. $5\pi a^2/4$ **7.** $e^{4k\pi}$

8. 3.1 **9.** $3\pi a^2/8$

Exercise 2D

1,2.

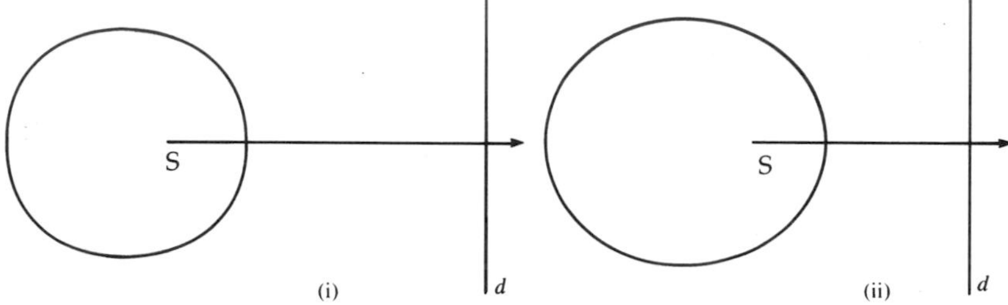

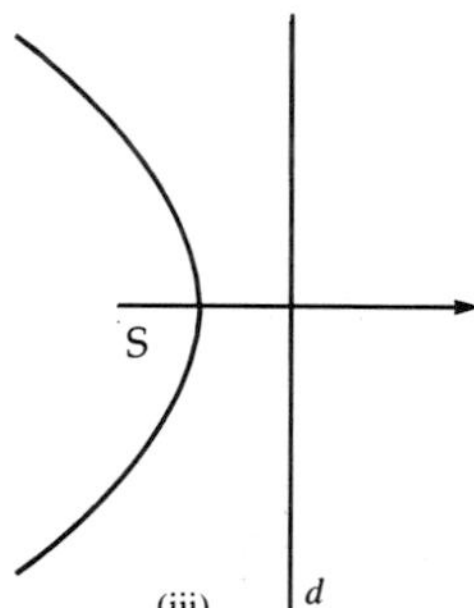

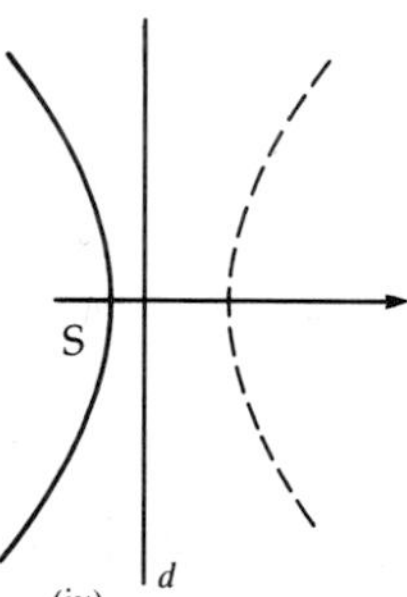

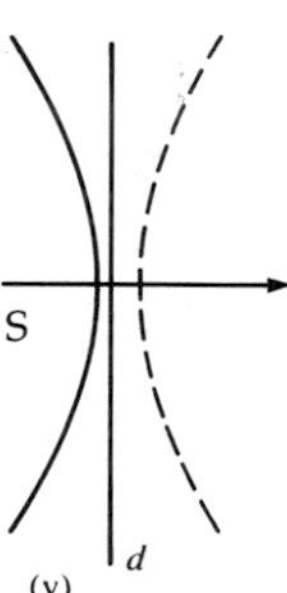

3.

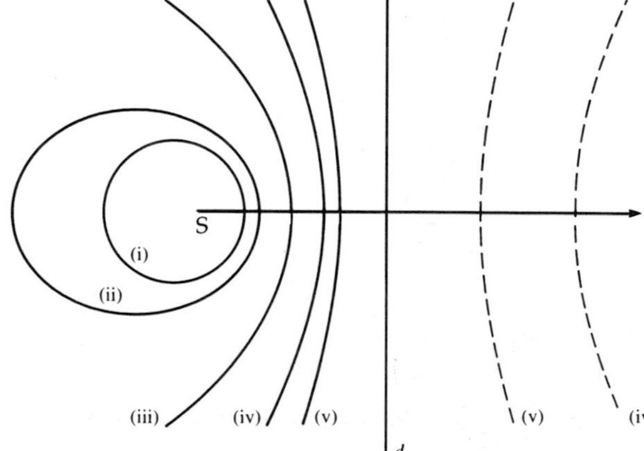

4. (i) Circle (ii) almost a pair of straight lines

5. (i) $(2a, \pm 2\pi/3)$ (ii)

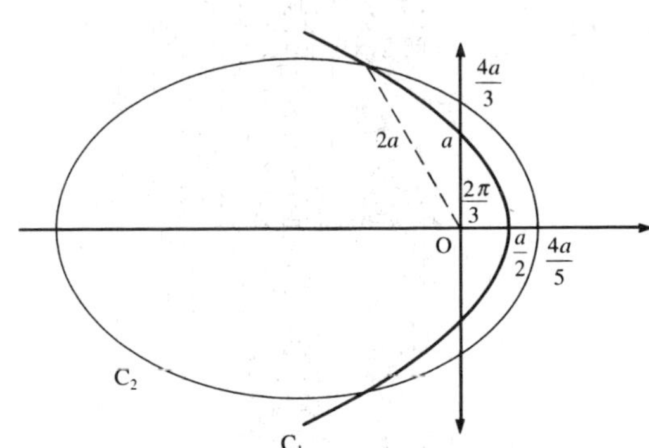

Investigation page 36

(ii) (a)

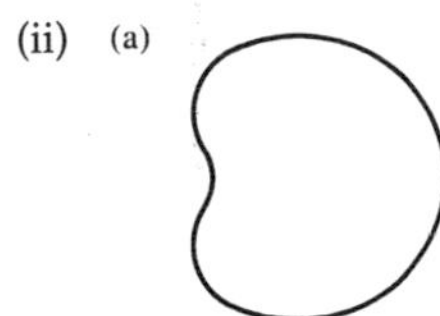

$k > 2a$

(b)

$k < 2a$

(iv) $\pi(2a^2 + k^2)$

Exercise 3A

1. (i) $-\frac{x^2}{y^2}, -\frac{2x}{y^5}$ (ii) $\frac{y-1}{1-x}, \frac{2(y-1)}{(1-x)^2}$ (iii) $-\frac{y+7}{x-3}, \frac{2(y+7)}{(x-3)^2}$

(iv) $\frac{y^2-2}{3-2xy}, -\frac{2(y^2-2)(8x+3y)}{(3-2xy)^3}$ (v) $-\frac{\cos x}{\cos y}, \frac{\sin x \cos^2 y + \sin y \cos^2 x}{\cos^3 y}$

(vi) $(\ln k)y$, $(\ln k)^2 y$ (vii) $y(1 + \ln x)$, $y(1 + \ln x)^2 + \frac{y}{x}$

2. $-\frac{7}{6}$ **3.** $(2, 4)$, $-\frac{1}{3}$, maximum; $(-2, -4)$, $\frac{1}{3}$, minimum **4.** $(8, 4)$, $(-8, -4)$

5. $(\sqrt{(\sqrt{5} - 2)}, -1/\sqrt{(\sqrt{5} - 2)})$, $(-\sqrt{(\sqrt{5} - 2)}, 1/\sqrt{(\sqrt{5} - 2)})$; $(-1, -1)$, $(1, 1)$

6.

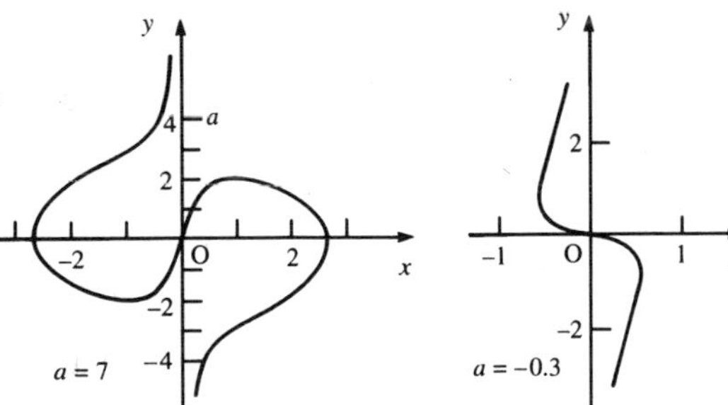

9. $\frac{s'tuvw + st'uvw - stu'vw - stuv'w - stuvw'}{(uvw)^2}$

10. (i) $2(2x-3)^2(x+4)^3(7x+6)$ (ii) $\frac{76-15x}{(5x+2)^3}$ (iii) $4x(7x^2 + 19)(2x^2 + 5)^2(x^2 + 3)^3$

(iv) $\frac{-2x(x^2+1)(x^2+9)}{(x^2-3)^4}$ (v) $\frac{2(x^2+3)^2(x^3-x^2-18x+6)}{(x-1)^3(2x+1)^4}$ (vi) $\frac{-x(x-1)(x^2+x+4)}{2(x^2+1)^{\frac{1}{2}}(x^3+2)^{\frac{3}{2}}}$

11. $\frac{p(b-cT)}{T^2}$ **12.** $\frac{2\cos B + \cos C}{\cos A - \cos C}$ **13.** $-\frac{x-1}{y-1}$ **14.** $\frac{(\gamma+1)V}{\gamma^2}$ **15.** $p = \frac{a}{27b^2}, V = 3b, T = \frac{8a}{27bR}$

Exercise 3B

1.

	arcsine	arccosine	arctangent
Domain	$-1 \leqslant x \leqslant 1$	$-1 \leqslant x \leqslant 1$	all real numbers
Range	$-\frac{\pi}{2} \leqslant y \leqslant \frac{\pi}{2}$	$0 \leqslant y \leqslant \pi$	$-\frac{\pi}{2} \leqslant y \leqslant \frac{\pi}{2}$

3. $\arccos x + \arccos(-x) = \pi$ **4.** $-\frac{\pi}{2} \leqslant x \leqslant \frac{\pi}{2}$

6. (i) $\frac{1}{\sqrt{1-x^2}}$ (ii) $\frac{5}{\sqrt{1-25x^2}}$ (iii) $\frac{6}{4+9x^2}$ (iv) $\frac{-3}{1+(2-3x)^2}$

7. (i) $\frac{2}{\sqrt{1-4x^2}}$ (ii) $\frac{5}{1+25x^2}$ (iii) $\frac{6x}{\sqrt{1-9x^4}}$ (iv) $-\frac{2}{\sqrt{1-4x^2}}$

(v) $\frac{e^x}{1+e^{2x}}$ (vi) $\frac{-2x}{1+(1-x^2)^2}$ (vii) $-\frac{10x}{\sqrt{1-(5x^2-2)^2}}$ (viii) $\frac{1}{2\sqrt{x(1-x)}}$

8. $\arcsin\frac{x}{\sqrt{2}} - \frac{\pi}{4}$ **9.** $\frac{1}{\sqrt{1-x^2}}, -\frac{1}{\sqrt{1-x^2}}; c_2 - c_1 = \frac{\pi}{2}$

10. $n\pi$ or $2n\pi \pm \frac{\pi}{3}$ **11.** $2n\pi - \frac{\pi}{4}$ **12.** $2n\pi \pm \frac{\pi}{3} + \arcsin\frac{4}{5}$

13. $n\pi$ or $n\pi \pm \arctan\sqrt{\frac{1}{2}}$ **14.** $4n\pi$ or $4n\pi \pm \frac{4\pi}{3}$ **15.** $(2n+1)\pi$ or $2n\pi + 2\arcsin\frac{1}{\sqrt{5}}$

16.

	arcsecant	arccosecant	arccotangent
Domain	$x \leqslant -1$ or $x \geqslant 1$	$x \leqslant -1$ or $x \geqslant 1$	all real numbers
Range	$0 \leqslant y \leqslant \pi, y \neq \frac{\pi}{2}$	$-\frac{\pi}{2} \leqslant y \leqslant \frac{\pi}{2}, y \neq 0$	$-\frac{\pi}{2} < y \leqslant \frac{\pi}{2}, y \neq 0$

17. (ii) (a) $-\frac{1}{|x|\sqrt{x(x^2-1)}}$ (b) $-\frac{1}{1+x^2}$

18. (i) $\frac{\pi}{2}$ provided $x \leqslant -1$ or $x \geqslant 1$ (ii) $-\frac{\pi}{2}$ if $x < 0$, $\frac{\pi}{2}$ if $x > 0$

Exercise 3C

1. $\frac{1}{5}\arctan\frac{x}{5} + c$ **2.** $\arcsin\frac{x}{6} + c$ **3.** $\frac{5}{6}\arctan\frac{x}{6} + c$

4. $\frac{2}{5}\arctan\frac{2x}{5} + c$ **5.** $\frac{1}{2}\arcsin\frac{2x}{3} + c$ **6.** $\frac{7}{\sqrt{3}}\arcsin\frac{\sqrt{3}x}{\sqrt{5}} + c$

7. $\frac{\pi}{12}$ **8.** $\frac{\pi}{4}$ **9.** $\frac{7\pi}{36}$ **10.** $\frac{\pi}{12}$ **11.** $\frac{\pi}{2\sqrt{6}}$ **12.** $\frac{\pi}{12\sqrt{10}}$

Exercise 3D

1. (i) $\frac{1}{2}\arctan\frac{x+2}{2} + c$ (ii) $7\arcsin\frac{x-2}{3} + c$ (iii) $\frac{\sqrt{3}}{\sqrt{2}}\arctan\frac{\sqrt{2}x}{\sqrt{3}} + c$

(iv) $\frac{1}{2}\arctan\frac{3x+1}{2} + c$ (v) $\arcsin\frac{x-1}{2} + c$ (vi) $\frac{7}{2}\arcsin\frac{2x+1}{2} + c$

2. (i) $x \arcsin x + \sqrt{1-x^2} +$

(ii) (a) $x \arccos x - \sqrt{1-x^2} +$ (b) $x \arctan x - \frac{1}{2}\ln(1+x^2)+c$ (c) $x \operatorname{arccot} x + \frac{1}{2}\ln(1+x^2)+c$

3. (i) $\frac{1}{2}a^2 \arcsin\frac{b}{a} + \frac{1}{2}b\sqrt{a^2-b^2}$ (ii) area of sector + area of triangle

4. (i) $\frac{1}{2}\arctan\frac{x-3}{2}+c$ (ii) $\frac{1}{2}\arcsin\frac{2x+3}{4}+c$ (iii) $\frac{1}{4}\arctan\frac{2x+5}{2}+c$

(iv) $-\frac{1}{x-3}+c$ (v) $\frac{1}{3}\arcsin\frac{3x+2}{3}+c$

5. (i) $\frac{1}{2}\ln(x^2+1)+\arctan x+c$ (ii) $\ln\frac{(x+1)^2}{x^2+1}+2\arctan x+c$

(iii) $\sqrt{1-x} + \arcsin x + c$ (iv) $\ln\frac{|x+1|}{\sqrt{x^2+1}}+2\arctan x+c$

6. (i) $\frac{\pi}{3}$ (ii) $\frac{1}{2}$ (ln58 + arctan $\frac{5}{2}$) $-\frac{\pi}{8} \approx 2.233$

7. $\frac{1}{x\sqrt{x^2-1}}$; $\frac{1}{a}\operatorname{arc\,sec}\frac{x}{a}+c$

Exercise 3E

1. $x+\frac{1}{3}x^3$ **2.** $1+\frac{1}{2}x^2+\frac{5}{24}x^4$ **3.** $x-\frac{1}{2}x^2+\frac{1}{6}x^3+\frac{1}{12}x^4$ **4.** $3x-\frac{9}{2}x^3$ **5.** $1-2x^2+\frac{2}{3}x^4$

6. $x+\frac{1}{6}x^3$ **7.** $x^2-\frac{1}{3}x^4$ **8.** $1+x+\frac{1}{2}x^2-\frac{1}{8}x^4$ **9.** $e+ex+\frac{1}{2}ex^2+\frac{1}{6}ex^3+\frac{1}{24}ex^4$

11. (iv) 3.14159

12. (i) $1-\frac{1}{2}x^2+\frac{1}{8}x^4-\frac{1}{48}x^6$ (ii) $1-\frac{1}{2}x^2+\frac{1}{8}x^4-\frac{1}{48}x^6+\frac{1}{384}x^8$; 0.8555 ± 0.00015

13. (i) 0.6456 (ii) 0.6911 (iii) $2x+\frac{2x^3}{3}+\frac{2x^5}{5}$; $x=\frac{1}{3}$ gives $\ln 2 \approx 0.6930$

14. (i) $a_0 = 2$ (ii) $2+x-x^2-\frac{x^3}{3}+\frac{x^4}{4}+\frac{x^5}{15}-\frac{x^6}{24}$

(iii)

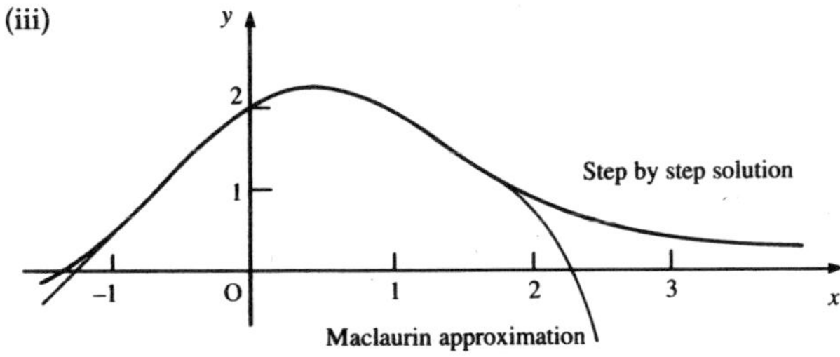

15. $\frac{k(k-1)(k-2)(k-3)\ldots(k-r+1)}{r!}\left(\frac{x}{k}\right)^r$

16. (i) $x+x^2+\frac{x^3}{3}-\frac{x^5}{30}-\frac{x^6}{90}$

(ii) (a) $\left(1+x+\frac{x^2}{2}+\frac{x^3}{6}\right)\left(x-\frac{x^3}{6}\right)=x+x^2+\frac{x^3}{3}-\frac{x^5}{12}-\frac{x^6}{36}$, agrees with series for $e^x \sin x$ as far as term in x^4;

(b) $1+x-\frac{x^3}{3}$ (the term of the product in x^4 is not valid)

17. (ii) $\ln 2 + x$

Exercise 4A

1. $(1-j)/\sqrt{2}$ **2.** $-(1+\sqrt{3}j)/2$ **3.** $-(\sqrt{3}+j)/2$ **4.** $(-1+j)/\sqrt{2}$ **5.** $-8+8\sqrt{3}j=-8+13.856j$
6. $-1024-1024j$ **7.** $-0.078+0.997j$ **8.** -46656 **9.** $\cos 8\alpha - j\sin 8\alpha$
10. $\cos 2\beta - j\sin 2\beta$ **11.** $\cos^{10}\gamma(\cos 10\gamma + j\sin 10\gamma)$ **12.** $(\cos 4\delta - j\sin 4\delta)/16\cos^4\delta$

Exercise 4B

1. $\dfrac{4t-4t^3}{1-6t^2+t^4}$, where $t=\tan\theta$

2. (i) $c^3-3cs^2=4c^3-3c$ (ii) $3c^2s-s^3=3s-4s^3$ (iii) $\dfrac{3t-t^3}{1-3t^2}$ where $c=\cos\theta$, $s=\sin\theta$ and $t=\tan\theta$

3. $32c^6-48c^4+18c^2-1$, $32c^5-32c^3+6c$, where $c=\cos\theta$

4. $\dfrac{{}^nC_1t-{}^nC_3t^3+\ldots}{1-{}^nC_2t^2+{}^nC_4t^4-\ldots}$ **5.** $(\cos 4\theta+4\cos 2\theta+3)/8$

6. $(\sin 5\theta-5\sin 3\theta+10\sin\theta)/16$ **7.** $(-\cos 6\theta+6\cos 4\theta-15\cos 2\theta+10)/32$
8. $(\cos 7\theta-\cos 5\theta-3\cos 3\theta+3\cos\theta)/64$ **9.** $(-\sin 7\theta-\sin 5\theta+3\sin 3\theta+3\sin\theta)/64$

11. $-\dfrac{1}{192}\sin 6\theta+\dfrac{3}{64}\sin 4\theta-\dfrac{15}{64}\sin 2\theta+\dfrac{5}{16}\theta+k$ **12.** $\dfrac{2}{35}$ **13.** $\dfrac{4}{35}$ **14.** $\dfrac{\sin 2n\theta}{2\sin\theta}$

15. $\dfrac{\sin\theta+\sin(n-1)\theta-\sin n\theta}{2-2\cos v}$ **16.** (i) $\sqrt{3}(\cos\frac{\pi}{6}+j\sin\frac{\pi}{6})$ (iii) $3^{n/2}\sin\dfrac{n\pi}{6}$

17. $2^n\cos^n\dfrac{\beta}{2}\sin\left(\alpha+\dfrac{n\beta}{2}\right)$ **18.** 0

Exercise 4C

3. $-\alpha$, $\pm\alpha\omega$, $\pm\alpha\omega^2$ **5.** (iv) If and only if m and n have no common factor

6. $j/2$, $(\pm\sqrt{3}+j)/2$ **7.** $\cos\dfrac{k\pi}{3}+j\sin\dfrac{k\pi}{3}$, $k=1, 2, 3, 4, 5$

9. $\dfrac{\cos\frac{2k\pi}{n}}{1-\sin\frac{2k\pi}{n}}$, $k=0, 1, 2, \ldots, n-1$ (excluding $k=3n/4$ if n is a multiple of 4)

10. $\cot\dfrac{(2k+1)\pi}{2n}$, $k=0, 1, 2, \ldots, n-1$

Exercise 4D

1. $\pm(0.90+2.79j)$ **2.** $\pm1\pm j$ **3.** $-119-120j$, $-3+2j$, $-2-3j$, $3-2j$

4.

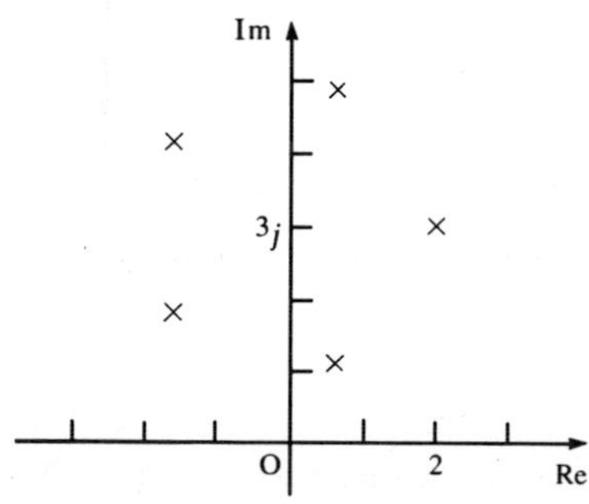

5. $(z+1-3j)^7=2187$ **7.** Regular n-gon with one vertex at O **9.** $3na^2/2$

11. $8\left(\cos\dfrac{\pi}{10}+j\sin\dfrac{\pi}{10}\right)$ **12.** $\dfrac{1}{49}\left(\cos\left(-\dfrac{\pi}{4}\right)+j\sin\left(-\dfrac{\pi}{4}\right)\right)$ **13.** $\dfrac{1}{2187}\left(\cos\left(\dfrac{7\pi}{12}\right)+j\sin\left(\dfrac{7\pi}{12}\right)\right)$

14. (i) $(-1+j)/\sqrt{2}$, $(1-j)/\sqrt{2}$, $(j^{1/2})^3=j^{3/2}$ (ii) both $(1+\sqrt{3}j)/2$ (iii) $-\pi < m\arg w \leqslant \pi$
15. No (replace $\sqrt{-1}$ by j and ∞ by n, then let $n\to\infty$)

Exercise 4E

2.

3.

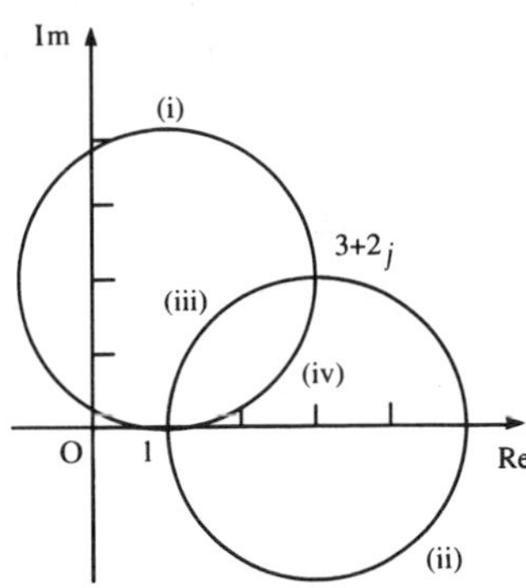

4. Converse not true **5.** $ae^* + bf^* + cd^* = af^* + bd^* + ce^*$

7. (ii) $3 - 2\sqrt{3} + (5 + \sqrt{3})j,\ 3 + 2\sqrt{3} + (5 - \sqrt{3})j$ **12.** (ii) Result still true

13. If and only if P, Q, R, S are as in Question 12 (i.e. isosceles triangles with right angles at P, Q, R, S)

Exercise 4F

1. -1 **2.** $(1 + j)/\sqrt{2}$ **3.** $-1.209 + 0.698j$ **4.** $-13.129 + 15.201j$

5. $3 + 2k\pi j$ **6.** $-4 + (2k - \frac{1}{3})\pi j$ **7.** All z

9. (i) $\cos\theta + j\sin\theta,\ \cos n\theta + j\sin n\theta,\ \cos n\theta - j\sin n\theta$

(iv) $C = \dfrac{2\cos\theta}{5 - 4\cos 2\theta},\quad S = \dfrac{6\sin\theta}{5 - 4\cos 2\theta}$

11. $\dot{z} = (\dot{r} + jr\dot{\theta})e^{j\theta},\ \ddot{z} = (\ddot{r} - r\dot{\theta}^2 + j(2\dot{r}\dot{\theta} + r\ddot{\theta}))e^{j\theta}$, where the dot shows differentiation with respect to t

Components:

	radial	**transverse**
velocity	$\dot{r}$	$r\dot{\theta}$
acceleration	$\ddot{r} - r\dot{\theta}^2$	$2\dot{r}\dot{\theta} + r\ddot{\theta}$

12. $C = e^{3x}(3\cos 2x + 2\sin 2x)/13 + c,\quad S = e^{3x}(-2\cos 2x + 3\sin 2x)/13 + c'$

13. $\dfrac{e^{ax}(a\cos bx + b\sin bx)}{a^2 + b^2} + c,\quad \dfrac{e^{ax}(-b\cos bx + a\sin bx)}{a^2 + b^2} + c'$

15. $\ln|z| + j(\arg z + 2k\pi)$

16. (i) $j\pi$ (ii) $j\pi/2$ (iii) $\ln 5 - j\pi/2$ (iv) $\ln 5 + j\arctan\frac{3}{4} \approx 1.609 + 0.644j$

17. $e^{-\pi/2} \approx 0.208$ **18.** (i) 1 (ii) 0.043 (iii) $2.808 - 1.318j$ (iv) $0.129 + 0.034j$

19. Moves along the real axis from $e^{-\pi}$ to e^{π}

Exercise 5A

2. (i) Vertex (0, 0), focus (3, 0), axis $y = 0$, directrix $x = -3$ (ii) $(0, 0), (-\frac{1}{4}, 0), y = 0, x = \frac{1}{4}$

(iii) $(3, 2), (4, 2), y = 2, x = 2$ (iv) $(0, 0), (0, \frac{3}{2}), x = 0, y = -\frac{3}{2}$

(v) $(-2, -5), (0, -5), y = -5, x = -4$ (vi) $(-3, 5), (-3, 4\frac{1}{2}), x = -3, 2y = 11$

3. (i) $(y - 7)^2 = 16(x - 1)$ (ii) $(y - 2)^2 = -40(x + 4)$ (iii) $(x - 4)^2 = 32(y + 1)$

7. $8a^2/3$ **9.** (i) $(apq, a(p + q))$

13. (i) $y^2 = 2ax$ (ii) $y^2 = a(x - a)$ (iii) $y^2 = 2a(x - a)$

15. (i) $y + rx = a(p + q + pqr)$ (ii) $(-a, a(p + q + r + pqr))$

Activity (page 96)

$b = \frac{\ell}{\sqrt{1-e^2}}, \ \frac{a}{b} = \frac{1}{\sqrt{1-e^2}}$

Activity (page 97)

(i) $\frac{4}{5}$ (ii) $\frac{7}{25}$ (iii) $\sqrt{0.9999} \approx 0.99995$

Activity (page 97)

(i) $b = 75$ cm, $\ell = 56$ cm (ii) $a = 134$ cm, $\ell = 75$ cm (iii) $a = 180$ cm, $b = 134$ cm

Exercise 5B

1. $e = \frac{3}{5}$ **2.** 1.14×10^{10} km **4.** 0.937 **5.** 800 km **6.** $\frac{8}{481} \approx 0.0166$

7. $\frac{(x-1)^2}{11} + \frac{(y-4)^2}{2} = 1$; centre (1, 4); foci (–2, 4), (4, 4); directrices $x = \frac{14}{3}, x = -\frac{8}{3}$

8. $\frac{(x+2)^2}{8} + \frac{(y-2)^2}{4} = 1$; centre (–2, 2); foci (–4, 2), (0, 2); directrices $x = -6$, $x = 2$

10. At the midpoint of the ladder **12.** (ii) $\left(\frac{-a^2mc}{a^2m^2+b^2}, \frac{b^2c}{a^2m^2+b^2}\right)$ (iv) $\frac{a}{b}\tan\theta$

14. $y = -x + 3$, (2, 1); $y = 11x + 27$, $\left(-\frac{22}{9}, \frac{1}{9}\right)$

15. (ii) $\frac{X^2}{a^2} + \frac{Y^2}{b^2} < 1$, (X, Y) inside the ellipse (iii) $X^2 + Y^2 = a^2 + b^2$

Exercise 5C

3. $\frac{x^2}{16} - \frac{y^2}{48} = 1$ **4.** $9x^2 - y^2 = 20$, $9x - 2y = 10$, $2x + 9y = 40$

5. (i) $(b^2 - a^2m^2)x^2 - 2a^2mcx - a^2(b^2 + c^2) = 0$, $\frac{2a^2mc}{b^2 - a^2m^2}$;

(ii) $(b^2 - a^2m^2)x^2 - 2a^2mcx - a^2c^2 = 0$, $\frac{2a^2mc}{b^2 - a^2m^2}$

8. $a^2m^2 = b^2 + c^2 \Rightarrow y = mx + c$ touches ***H*** or is an asymptote

10. The 'circle' shrinks to the point O; the 'perpendicular tangents' are the asymptotes

11. (v) (at, bt), $\left(\frac{a}{t}, -\frac{b}{t}\right)$

Exercise 5D

2. $(\sqrt{2}c, \sqrt{2}c)$, $(-\sqrt{2}c, -\sqrt{2}c)$; $x + y = \pm\sqrt{2}c$ **3.** (3, 2), (–3, –2)

6. $t^3x - ty + c(1 - t^4) = 0$; $R(-ct, -c/t)$ **7.** $2x = h$, $2y = k$

Exercise 5E

4. (i) Ellipse, foci A_1, A_2;

(ii) hyperbola, foci A_1, A_2 with one branch through the intersections of the circles

5. Points to the right of the right hand branch of $3x^2 - y^2 = 7500$

8. (iii) Concentric circles and straight lines through their centre

Exercise 5F

12. Hint: at the viewing point P the rays of light from the ellipse form a circular cone. In the plane section perpendicular to the ellipse the Dandelin sphere gives a variable circle touching AA′ at S (or S′). Prove that PA′ – PA = S′S.

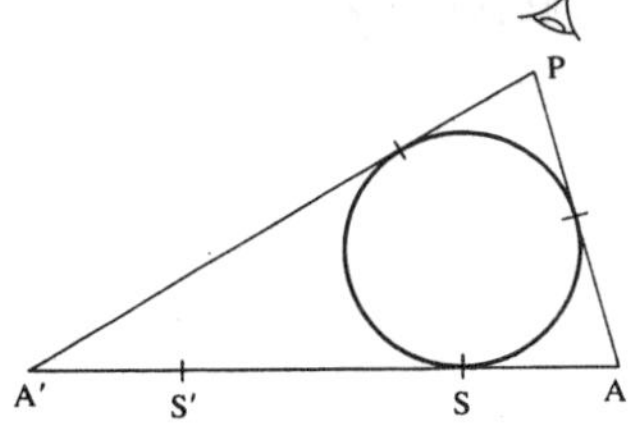

Exercise 6A

1. $\cosh A - \cosh B = 2\sinh\frac{A+B}{2}\sinh\frac{A-B}{2}$, $\sinh A + \sinh B = 2\sinh\frac{A+B}{2}\cosh\frac{A-B}{2}$,

$\sinh A - \sinh B = 2\cosh\frac{A+B}{2}\sinh\frac{A-B}{2}$

2. $\sinh 3u = 3\sinh u + 4\sinh^3 u$, $\cosh 3u = 4\cosh^3 u - 3\cosh u$

3. (i) (a) $-\ln 3$, (b) $\ln\frac{3}{4}, \ln 2$, (c) no solution (ii) $a+b$, $a-b$, c all have the same sign and $b^2 + c^2 > a^2$

4. $x = \ln 3, y = \ln 2$ **5.** 1.62 m, 22.3°

7. (i) $4\cosh 4x$ (ii) $2x\sinh(x^2)$ (iii) $2\cosh x\sinh x$ (iv) $\cos x\cosh x - \sin x\sinh x$

(v) $\frac{1}{2}\left(1+\frac{1}{x^2}\right)$ (vi) $5e^{10x}$ (vii) $3(1+x)^2\cosh^2 3x(\cosh 3x + 3(1+x)\sinh 3x)$ (viii) 1

8. $\frac{1}{2}(\cosh 2x + 1)$, $\frac{1}{2}(\cosh 2x - 1)$; $\frac{1}{4}\sinh 2x + \frac{1}{2}x + c$, $\frac{1}{4}\sinh 2x - \frac{1}{2}x + c$

9. (i) $\frac{1}{3}\cosh 3x + c$ (ii) $\frac{1}{2}\sinh(1+x^2) + c$ (iii) $x\cosh x - \sinh x + c$

(iv) $\sinh x + \frac{1}{3}\sinh^3 x + c$ (v) $\frac{1}{4}x\sinh 2x - \frac{1}{8}\cosh 2x - \frac{1}{4}x^2 + c$ (vi) $\frac{1}{18}e^{9x} - \frac{1}{2}e^{-x} + c$

(vii) $\frac{1}{5}\cosh^5 x - \frac{1}{3}\cosh^3 x + c$ (viii) $\frac{1}{28}\cosh 14x + \frac{1}{4}\cosh 2x + c$

10. $\cosh x = 1+\frac{x^2}{2!}+\frac{x^4}{4!}+\frac{x^6}{6!}+\ldots$, $\sinh x = x+\frac{x^3}{3!}+\frac{x^5}{5!}+\frac{x^7}{7!}+\ldots$, both valid for all x; 6 terms

12. $(\cosh x - \sinh x)^n = \cosh nx - \sinh nx$;
$\cosh 5x = 16\cosh^5 x - 20\cosh^3 x + 5\cosh x$, $\sinh 5x = 16\sinh^5 x + 20\sinh^3 x + 5\sinh x$

Exercise 6B

1. (i)

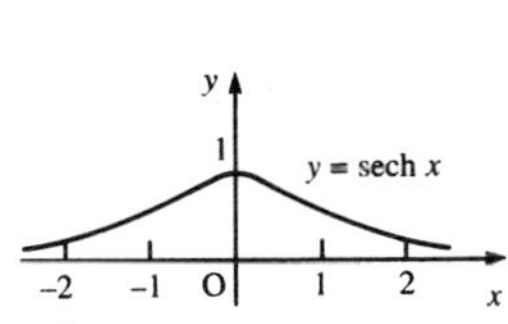

$y = 0$

(ii)

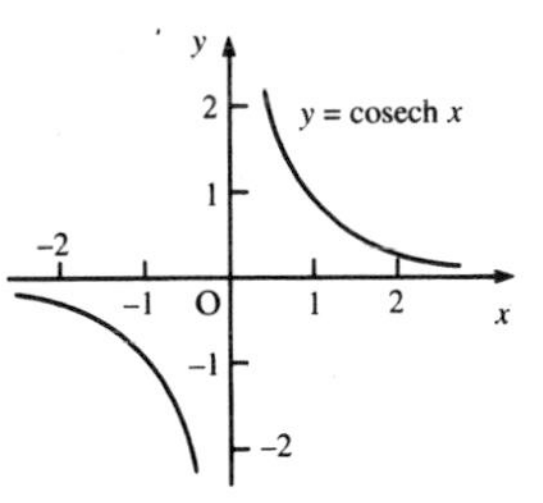

$x = 0,\ y = 0$

(iii)

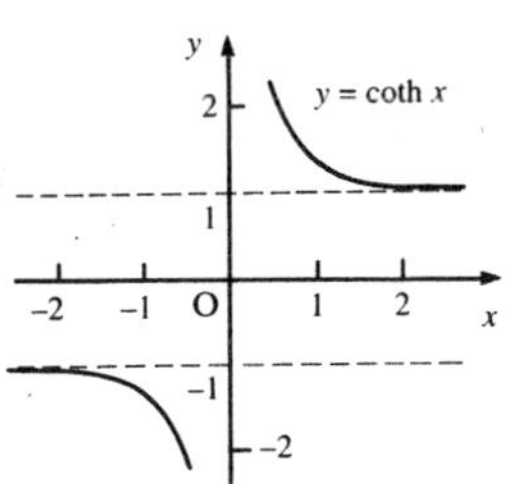

$x = 0,\ y = \pm 1$

3. (i) $\pm\frac{1}{2}\ln 3$, (ii) $0, \ln 7$, (iii) $0, \frac{1}{2}\ln 2$

4. $p = \frac{1}{2}\ln(2+\sqrt{5})$, $q = \ln(1+\sqrt{2})$

(i) $\tanh x < \sinh x < \operatorname{sech} x < \cosh x < \operatorname{cosech} x < \coth x$
(ii) $\tanh x < \operatorname{sech} x < \sinh x < \operatorname{cosech} x < \cosh x < \coth x$

7. (i) $-\operatorname{sech} x\tanh x$ (ii) $-\operatorname{cosech} x\coth x$ (iii) $-\operatorname{cosech}^2 x$ (iv) $\operatorname{sech} x\operatorname{cosech} x$

8. (i) $\ln(\cosh x) + c$ (ii) $\ln|\sinh x| + c$ (iii) $2\arctan(e^x) + c$ (iv) $\ln\left|\frac{e^x-1}{e^x+1}\right| + c$

9. $x - \frac{1}{3}x^3 + \frac{2}{15}x^5$ **10.** $-2.68 < x < 2.68$

Exercise 6C

3.

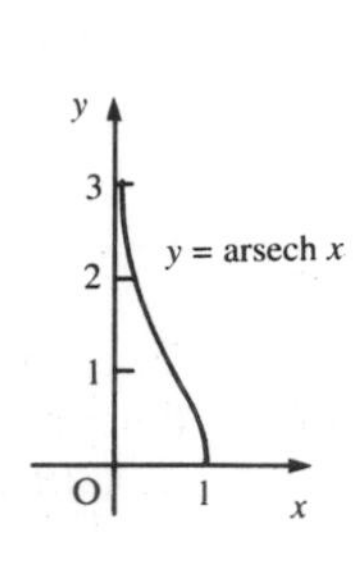

$0 < x \leqslant 1, y \geqslant 0$

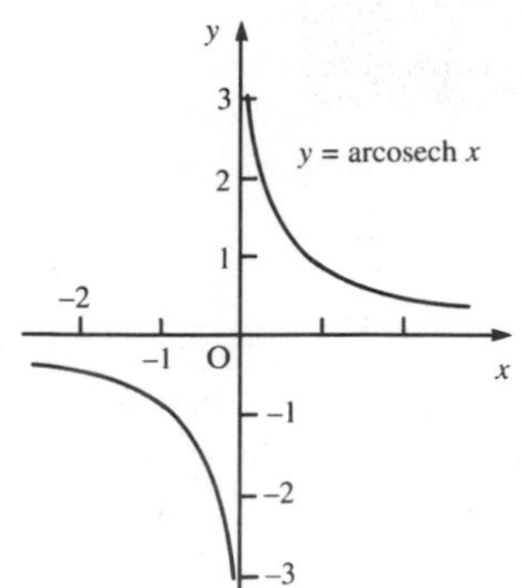

$x \neq 0, y \neq 0$

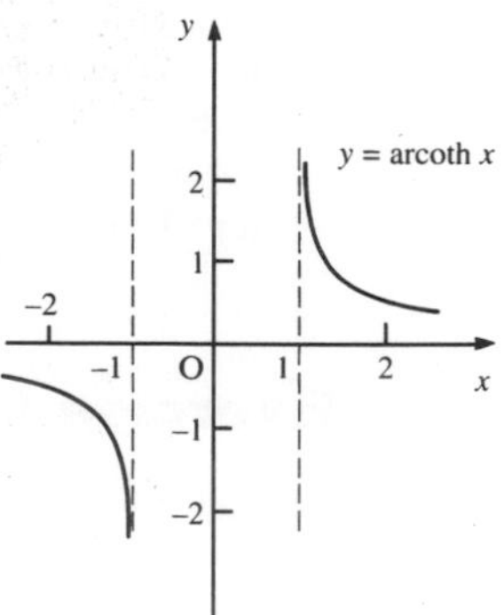

$x < -1$ or $x > 1, y \neq 0$

4. (i) $\dfrac{3}{\sqrt{9x^2+1}}$ (ii) $\dfrac{2x}{\sqrt{x^4-1}}$ (iii) $\operatorname{sech} x$ (iv) $\sec x$ (v) $-\dfrac{1}{x\sqrt{1-x^2}}$

5. (i) $x \operatorname{arcosh} x - \sqrt{x^2-1}+c$ (ii) $x \operatorname{arsinh} x - \sqrt{x^2+1}+c$ (iii) $x \operatorname{artanh} x + \frac{1}{2}\ln(1-x^2)+c$

6. (i) $\operatorname{arsinh}(x/2)+c$ (ii) $\operatorname{arcosh}(x/3)+c$ (iii) $\arcsin(x/3)+c$ (iv) $\frac{1}{6}\operatorname{arsinh}(3x/2)+c$

(v) $\operatorname{arsinh}(\frac{1}{2}x-1)+c$ (vi) $\operatorname{arcosh}(2x+1)+c$ (vii) $\frac{1}{3}\operatorname{arcosh}(x+\frac{1}{3})+c$ (viii) $\frac{1}{3}\operatorname{arcosh}(x^3)+c$

7. (i) 0.494 (ii) 0.322

9. (i) $\frac{1}{2}x\sqrt{a^2-x^2}+\frac{1}{2}a^2\arcsin\left(\frac{x}{a}\right)+c$ (ii) $\frac{1}{2}x\sqrt{a^2+x^2}+\frac{1}{2}a^2\operatorname{arsinh}\left(\frac{x}{a}\right)+c$

(iii) $\frac{1}{2}x\sqrt{x^2-a^2}-\frac{1}{2}a^2\operatorname{arcosh}\left(\frac{x}{a}\right)+c$

10. $\displaystyle\sum_{r=1}^{\infty}\frac{x^{2r-1}}{2r-1}, -1<x<1$ **11.** $\displaystyle\sum_{r=1}^{\infty}(-1)^{r-1}\frac{1^2.3^2.5^2...(2r-3)^2}{(2r-1)!}x^{2r-1}$

Exercise 6D

1. $-4.190 - 9.109j$; $-3.725 + 0.512j$ **2.** $\sinh x \cos y$, $\cosh x \sin y$

4. $\tan z = \dfrac{e^{2jz}-1}{j(e^{2jz}+1)}$, period π; $\tanh z = \dfrac{e^{2z}-1}{e^{2z}+1}$, period $j\pi$

5. (i) $\sin x \cosh y$, $\cos x \sinh y$ **6.** $j(2n+1)\pi$

7. $(2k-\frac{1}{2})\pi \pm j\operatorname{arcosh}(-h)$

9. The cosine curve $h=\cos x, y=0$ in the x-h plane, with ±cosh curves parallel to the y-h plane coming from its turning points: $h=\cosh y, x=2k\pi$ and $h=-\cosh y, x=(2k+1)\pi$

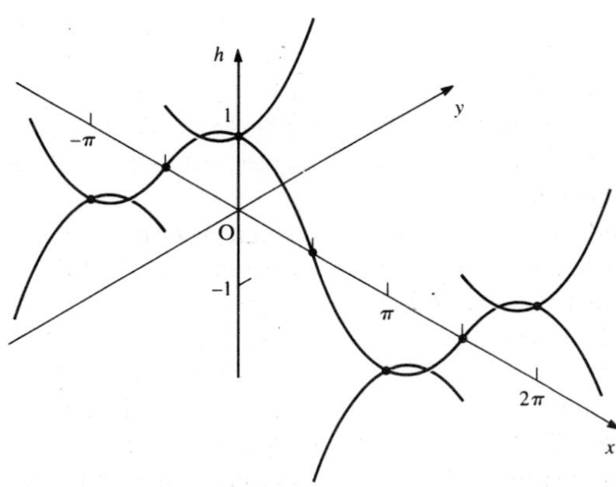

Index